Advances in Algebra Analysis and Topology

This book presents cutting-edge research, advanced techniques, and practical applications of Algebra Analysis and Topology. It offers in-depth insights, theoretical developments, and practical applications, showcasing the richness and interdisciplinary nature of algebra, analysis, and topology. The book fosters a deeper understanding of the fundamental principles while also highlighting the latest advancements and emerging trends in these disciplines. Readers are encouraged to apply the theoretical concepts and techniques to solve mathematical problems, engaging with the book's problem-solving approach. By combining theoretical foundations, practical applications, and interdisciplinary perspectives, this book aims to inspire new avenues of research and contribute to the ongoing development of these dynamic fields.

- Provides a comprehensive and accessible resource that covers a broad range of topics in algebra, analysis, and topology, understanding of the interconnections between these mathematical fields
- Encompasses both classical topics and cutting-edge research areas within algebra, analysis, and topology
- Covers foundational concepts, advanced theories, and their applications in diverse fields such as physics, computer science, engineering, and economics
- Offers sophisticated tools and methodologies to tackle complex problems and deepen the understanding of these disciplines
- Explores how algebra, analysis, and topology intersect with other fields of mathematics and how their concepts and techniques can be applied in related disciplines

It serves as a valuable reference for graduate students, researchers, and mathematicians seeking to deepen their knowledge and engage with the latest advancements in these fundamental branches of mathematics.

Advances in Algebra Analysis and Topology

Edited by
Sandeep Singh, Kalyan Chakraborty,
Baljinder Kour and Sandeep Kaur

CRC Press
Taylor & Francis Group
Boca Raton London New York

CRC Press is an imprint of the
Taylor & Francis Group, an **informa** business
A CHAPMAN & HALL BOOK

Designed cover image: ShutterStock

First edition published 2025
by CRC Press
2385 NW Executive Center Drive, Suite 320, Boca Raton FL 33431

and by CRC Press
4 Park Square, Milton Park, Abingdon, Oxon, OX14 4RN

CRC Press is an imprint of Taylor & Francis Group, LLC

Library of Congress Cataloging-in-Publication Data
Names: Singh, Sandeep (Mathematician), editor.
Title: Advances in algebra analysis and topology / edited by Sandeep Singh, Kalyan Chakraborty, Baljinder Kour, Sandeep Kaur.
Description: First edition. | Boca Raton : C&H/CRC Press, 2025. | Includes bibliographical references and index. | Summary: "This book presents cutting-edge research, advanced techniques, and practical applications of Algebra Analysis and Topology. It offers in-depth insights, theoretical developments, and practical applications, showcasing the richness and interdisciplinary nature of algebra, analysis, and topology. The book fosters a deeper understanding of the fundamental principles while also highlighting the latest advancements and emerging trends in these disciplines. Readers are encouraged to apply the theoretical concepts and techniques to solve mathematical problems, engaging with the book's problem-solving approach. By combining theoretical foundations, practical applications, and interdisciplinary perspectives, this book aims to inspire new avenues of research and contribute to the ongoing development of these dynamic fields. Provides a comprehensive and accessible resource that covers a broad range of topics in algebra, analysis, and topology, understanding of the interconnections between these mathematical fields Encompasses both classical topics and cutting-edge research areas within algebra, analysis, and topology Covers foundational concepts, advanced theories, and their applications in diverse fields such as physics, computer science, engineering, and economics Offers sophisticated tools and methodologies to tackle complex problems and deepen the understanding of these disciplines Explores how algebra, analysis, and topology intersect with other fields of mathematics and how their concepts and techniques can be applied in related disciplines It serves as a valuable reference for graduate students, researchers, and mathematicians seeking to deepen their knowledge and engage with the latest advancements in these fundamental branches of mathematics"-- Provided by publisher.
Identifiers: LCCN 2024029161 (print) | LCCN 2024029162 (ebook) | ISBN 9781032620336 (hbk) | ISBN 9781032634135 (pbk) | ISBN 9781032634142 (ebk)
Subjects: LCSH: Mathematical analysis. | Topology.
Classification: LCC QA300 .A336 2025 (print) | LCC QA300 (ebook) | DDC 515/.13--dc23/eng20240812
LC record available at https://lccn.loc.gov/2024029161
LC ebook record available at https://lccn.loc.gov/2024029162

ISBN: 978-1-032-62033-6 (hbk)
ISBN: 978-1-032-63413-5 (pbk)
ISBN: 978-1-032-63414-2 (ebk)

DOI: 10.1201/9781032634142

Typeset in Nimbus Roman
by KnowledgeWorks Global Ltd.

Publisher's note: This book has been prepared from camera-ready copy provided by the authors. This line should be placed below the DOI in CR ready title's copyright page

Dedication

Dedicated
to
God

Contents

Preface

It is with great pleasure that we present "Advances in Algebra Analysis and Topology" a comprehensive compilation that explores fundamental concepts and advanced topics in these foundational branches of mathematics. This book brings together a collection of cutting-edge research articles contributed by distinguished experts in the field. We aim to offer a diverse collection of topics, from foundational principles to cutting-edge research, presenting a valuable resource for mathematicians, researchers, and graduate students. Our editorial team, comprising Dr. Sandeep Singh, Dr. Kalyan Chakraborty, Dr. Baljinder Kour, and Dr. Sandeep Kaur, has endeavored to assemble a diverse selection of chapters that reflect the depth and breadth of current research. With thirteen chapters authored by twenty-seven leading authors in the field, this book aims to serve as a resource for both seasoned mathematicians and aspiring scholars alike. The idea behind this book stems from the recognition that algebra, analysis, and topology are not only pillars of mathematical theory but also crucial components of interdisciplinary applications. We have curated chapters authored by experts in their respective fields to provide both breadth and depth, fostering an appreciation for the interconnectedness and practical relevance of these disciplines. The scope of this book is intentionally broad, covering classical theories alongside modern advancements. We delve into abstract structures, functional analysis, operator theory, and their applications in diverse fields such as physics, computer science, and engineering. Each chapter is crafted to highlight the practical relevance and broad impact of mathematical concepts. From algebraic structures that underpin cryptography to topological methods in data analysis, the chapters illustrate how theoretical frameworks translate into real-world solutions. Moreover, we embrace diverse mathematical approaches—whether abstract or applied—to cater to readers with varying interests and backgrounds. This book serves as a bridge, connecting classical theories with contemporary research and providing a platform for exploration and discovery. Each chapter is complemented by a comprehensive list of references and further readings. These resources guide readers to explore specific topics in greater depth, providing avenues for continued learning and research. "Advances in Algebra Analysis and Topology" is designed to be more than a textbook—it is a gateway to the dynamic world of mathematical inquiry. Whether you are a graduate student embarking on a research journey or a seasoned mathematician seeking inspiration, we hope this book sparks curiosity and fosters appreciation for the elegance and utility of algebra, analysis, and topology. Lastly, we extend our heartfelt appreciation to all the contributors who have generously shared their expertise and knowledge. Their passion for mathematics is evident in the quality and depth of their contributions, enriching this volume immeasurably. As editors, it is our privilege to present "Advances in Algebra Analysis and Topology" to the mathematical community. It is our sincere hope that this book serves as a catalyst for new avenues of research and contributes to the ongoing evolution of these fundamental branches of mathematics.

Editor Bio

- **Dr. Sandeep Singh** is an Associate Professor with the Department of Mathematics at Akal University Talwandi Sabo, Bathinda, Punjab, India. He received his Ph.D. in Group Theory (Mathematics) from School of Mathematics, Thapar University, Patiala, India and M.Sc. degree in Mathematics from Punjabi University, Patiala, India. He was also a postdoctoral fellow at Department of Mathematics, Indian Institute of Technology Roorkee under programme SERB- NPDF (National Postdoctoral fellowship). His research interests include Group Theory, Automorphism Groups, Number Theory, Sum-set Problems, and Optimization Techniques. Besides holding an excellent academic record throughout, he has cleared the national level examinations NET-JRF conducted by UGC-CSIR, India. Also, He had been a recipient of CSIR junior and senior research fellowships for 2011-2013 and 2013-16, respectively. He has published more than 20 research papers in various journals of international repute and of different publishers. He has presented his research work at various international conferences and is the recipient of travel support by the Department of Science and Technology (DST), India. He has a teaching experience of around 8 years. He taught real analysis, ordinary differential equation, linear algebra, etc. at UG level and topics in algebra, optimization techniques, probability and statistics at PG Level. He has supervised more than 3 students for their M.Sc. dissertation work and currently supervising 2 Master's students at Akal University. He has also supervised 2 Ph.D. students and presently, 2 students are pursuing their Ph.D. under his supervision. He is a life member of Ramanujan Mathematical Society and Indian Mathematical Society and reviewer of American Mathematical Society.
- **Prof. Kalyan Chakraborty** holds a D. Phil. in Mathematics from the Harish-Chandra Research Institute. He is currently serving as a Professor and Head, Department of Mathematics, SRM University, Andhra Pradesh, India. His research prowess is evidenced by the publication of more than 58 research papers in prestigious journals, as well as contributions to book chapters and conference proceedings. Under his guidance, 6 Ph.D. scholars have completed their research, while currently, 5 Ph.D. scholars are actively pursuing their studies. Additionally, he has mentored 4 postdoctoral students, who continue to benefit from his mentorship. Prof. Chakraborty's wide-ranging research encompasses various topics in Number Theory, including the average behavior of arithmetical functions, Hecke eigenforms, Jacobi forms of higher degree, modular forms, cubic equations in quadratic fields, and the number of real quadratic fields with class numbers divisible by 3.
- **Dr. Baljinder Kour** is an Assistant Professor of Mathematics in the Department of Mathematics, Akal University, Talwandi Sabo, Bathinda, Punjab, India. She obtained her Ph.D. Degree in Applied Mathematics from the Central University of Punjab, Bathinda, India in 2021. She has cracked national level Lectureship and Junior research fellowship exam three times with all india rank 12, 32, 62 in 2015 and 2016. She graduated with an M.Sc. Mathematics from Kurukshetra University in 2013. She obtained a B.Ed. Educational Administration from Kurukshetra University in 2014. Dr. Kour serves as a Reviewer of internationally accredited journals and serves as an editorial board member of peer-reviewed journals. She has participated in many international conferences and workshops. She has been supervising Masters degree students at her university. She has twelve publications in referenced journals and two book chapters to her credit. She has a total of 69 Google citations and a 5-h-index for the past five years. Her research interests are Mathematical Modelling in educational issues, fractional calculus, symmetry analysis, real analysis, and operator theory.

- **Dr. Sandeep Kaur** is currently working as an Assistant Professor, Department of Mathematics and Statistics at Central University of Punjab, Punjab, India. He has cleared the national level examinations NET-JRF conducted by UGC-CSIR (2010) and GATE (2011), India. He has also received CSIR junior and senior research fellowships for 2011-2014 and 2014-15, respectively. He completed his Ph.D. from Department of Mathematics, Punjabi University, Patiala, under the guidance of Prof. (Dr.) Navpreet Singh Noorie. He has a teaching experience of ten semesters (Aug 2016-June 2022) in Central University of Punjab, Bathinda, India. He is also working on a project funded by TÜBİTAK-BİDEB, Turkey (2022). He has published my research in various international journals and a book chapter in "INSAC Academic Studies on Natural and Engineering Sciences". Dr. Kaur's research interests include General Topology, Fuzzy logic & Fuzzy set theory and soft set theory. He has supervised 27 students of M.Sc. Mathematics and M.Sc. Statistics for their project work at the Central University of Punjab, Bathinda, India, and delivered three seminars (offline) at Igdir University, Turkey on Induced mappings in Fuzzy and Soft Set theory and their applications. He has attended various ATM Schools workshops organized jointly by TIFR and IIT Bombay, India, and has presented my research in various international conferences.

List of Contributors

Mohammad Afajal Ansari
Aligarh Muslim University
India.

Md Shamim Akhter
Aligarh Muslim University
India.

Ashish Arora
IKG Punjab Technical University
India.

Mohammad Ashraf
Aligarh Muslim University
India.

Reeta Bhardwaj
Amity University
India.

Rohit Garg
Government Ripudaman College
India.

Bobin George
Pavanatma College
India

Nitakshi Goyal
Akal Degree College
India.

Erhan Guler
Bartin University
Turkey.

Jayakumar Jayaraman
Puducherry Technological University
India.

Jinny Ann John
Puducherry Technological University
India.

Jinta Jose
Viswajyothi College of Engineering and Technology
India

Tanupreet Kaur
Guru Nanak College
India.

Sandeep Kaur
Central University of Punjab
India.

Omer Kisi
Bartin University
Turkey.

Baljinder Kour
Akal University
India.

Tajender Kumar
University of Delhi
India.

Megha S. Pingale
Sandip University
India.

Renu P. Pathak
Sandip University
India.

Ritika
Pt. N. R. S. Government College
India.

Rajeshwari S
Bangalore Institute of Technology
India.

Sandeep Singh
Akal University
India.

Rajesh K. Thumbakara
Mar Athanasius College
India

Husna V
Presidency University
India.

Priyanka V
Presidency University
India.

1 Nonlinear Generalized Bi-skew Lie N-Derivations on $*$-Algebras

Mohammad Ashraf
Department of Mathematics, Aligarh Muslim University, Aligarh, India

Md Shamim Akhter
Department of Mathematics, Aligarh Muslim University, Aligarh, India

Mohammad Afajal Ansari
Department of Mathematics, Aligarh Muslim University, Aligarh, India

CONTENTS

Let $\mathscr{A}$ be a unital $*$-algebra with a nontrivial projection. In this chapter, it is proved that under some suitable assumptions, every nonlinear generalized bi-skew Lie n-derivation $\mathscr{G}:\mathscr{A}\to\mathscr{A}$ is of the form $\mathscr{G}(\mathscr{K})=\mathscr{V}\mathscr{K}+\mathscr{D}(\mathscr{K})$ for all $\mathscr{K}\in\mathscr{A}$, where $\mathscr{V}\in\mathscr{Z}(\mathscr{A})$ (the center of $\mathscr{A}$) and $\mathscr{D}:\mathscr{A}\to\mathscr{A}$ is an additive $*$-derivation. Moreover, it is shown that $\mathscr{G}$ is an additive generalized $*$-derivation. As an application, we give a description of nonlinear generalized bi-skew Lie n-derivations on certain special classes of unital $*$-algebras such as unital prime $*$-algebras, standard operator algebras and von Neumann algebras.

1.1 INTRODUCTION AND PRELIMINARIES

Let $\mathscr{A}$ be an associative $*$-algebra over the complex field $\mathbb{C}$. For any $\mathscr{K}_1,\mathscr{K}_2\in\mathscr{A}$, we denote the bi-skew Lie product of $\mathscr{K}_1$ and $\mathscr{K}_2$ by $[\mathscr{K}_1,\mathscr{K}_2]_\diamond=\mathscr{K}_1\mathscr{K}_2^*-\mathscr{K}_2\mathscr{K}_1^*$. Similar kinds of products based on involution $(*)$, such as the skew Lie product $([\mathscr{K}_1,\mathscr{K}_2]_\bullet=\mathscr{K}_1\mathscr{K}_2-\mathscr{K}_2\mathscr{K}_1^*)$, the skew Jordan product $(\mathscr{K}_1\bullet\mathscr{K}_2=\mathscr{K}_1\mathscr{K}_2+\mathscr{K}_2\mathscr{K}_1^*)$, the bi-skew Jordan product $(\mathscr{K}_1\diamond\mathscr{K}_2=\mathscr{K}_1\mathscr{K}_2^*+\mathscr{K}_2\mathscr{K}_1^*)$, etc., have been extensively studied on different kinds of $*$-algebras (see [2, 3, 5, 6, 7, 8, 9, 10, 11, 12, 14, 15, 13, 21, 22, 23, 24, 25, 26] and references therein). These new products play a significant role in certain research areas (see, for instances, [16, 17, 18, 19, 20]), and hence many authors have devoted themselves to the exploration of mappings involving these products across a variety of rings and algebras. Recall that an additive map $\mathscr{D}:\mathscr{A}\to\mathscr{A}$ is called an *additive $*$-derivation* if $\mathscr{D}(\mathscr{K}_1\mathscr{K}_2)=\mathscr{D}(\mathscr{K}_1)\mathscr{K}_2+\mathscr{K}_1\mathscr{D}(\mathscr{K}_2)$ and $\mathscr{D}(\mathscr{K}^*)=\mathscr{D}(\mathscr{K})^*$ hold for all

DOI: 10.1201/9781032634142-1

$\mathscr{K}, \mathscr{K}_1, \mathscr{K}_2 \in \mathscr{A}$. An additive map $\mathscr{G} : \mathscr{A} \to \mathscr{A}$ is said to be an *additive generalized $*$-derivation* if there exists an additive $*$-derivation $\mathscr{D}$ of $\mathscr{A}$ such that $\mathscr{G}(\mathscr{K}_1\mathscr{K}_2) = \mathscr{G}(\mathscr{K}_1)\mathscr{K}_2 + \mathscr{K}_1\mathscr{D}(\mathscr{K}_2)$ and $\mathscr{G}(\mathscr{K}^*) = \mathscr{G}(\mathscr{K})^*$ hold for all $\mathscr{K}, \mathscr{K}_1, \mathscr{K}_2 \in \mathscr{A}$. Suppose that $n \geq 2$ is a fixed integer and $\mathscr{K}_1, \mathscr{K}_2, \ldots, \mathscr{K}_n \in \mathscr{A}$. Let us define a sequence of polynomials as follows:

$$\begin{aligned}
\mathscr{Q}_1(\mathscr{K}_1) &= \mathscr{K}_1, \\
\mathscr{Q}_2(\mathscr{K}_1, \mathscr{K}_2) &= [\mathscr{Q}_1(\mathscr{K}_1), \mathscr{K}_2]_\diamond = [\mathscr{K}_1, \mathscr{K}_2]_\diamond, \\
&\vdots \qquad\qquad \vdots \\
\mathscr{Q}_n(\mathscr{K}_1, \mathscr{K}_2, \ldots, \mathscr{K}_n) &= [\mathscr{Q}_{n-1}(\mathscr{K}_1, \mathscr{K}_2, \ldots, \mathscr{K}_{n-1}), \mathscr{K}_n]_\diamond.
\end{aligned}$$

The element $\mathscr{Q}_n(\mathscr{K}_1, \mathscr{K}_2, \ldots, \mathscr{K}_n)$ is called the bi-skew Lie n-product of elements $\mathscr{K}_1, \mathscr{K}_2, \ldots, \mathscr{K}_n \in \mathscr{A}$. A map $\mathscr{D} : \mathscr{A} \to \mathscr{A}$ (not necessarily linear) is called a *nonlinear bi-skew Lie n-derivation* ($n \geq 2$) if

$$\mathscr{D}(\mathscr{Q}_n(\mathscr{K}_1, \mathscr{K}_2, \ldots, \mathscr{K}_n)) = \sum_{i=1}^{n} \mathscr{Q}_n(\mathscr{K}_1, \mathscr{K}_2, \ldots, \mathscr{K}_{i-1}, \mathscr{D}(\mathscr{K}_i), \mathscr{K}_{i+1}, \ldots, \mathscr{K}_n) \tag{1.1}$$

holds for all $\mathscr{K}_1, \mathscr{K}_2, \ldots, \mathscr{K}_n \in \mathscr{A}$. In particular, a nonlinear bi-skew Lie 2-derivation is called a *nonlinear bi-skew Lie derivation* and a nonlinear bi-skew Lie 3-derivation is said to be a *nonlinear bi-skew Lie triple derivation*. Recently, nonlinear bi-skew Lie (resp. Lie triple) derivations of von Neumann algebras have been studied in [9, 10].

A map $\mathscr{G} : \mathscr{A} \to \mathscr{A}$ (not necessarily linear) is said to be a *nonlinear generalized bi-skew Lie n-derivation* if there exists a nonlinear bi-skew Lie n-derivation $\mathscr{D} : \mathscr{A} \to \mathscr{A}$ such that

$$\mathscr{G}(\mathscr{Q}_n(\mathscr{K}_1, \mathscr{K}_2, \ldots, \mathscr{K}_n)) = \mathscr{Q}_n(\mathscr{G}(\mathscr{K}_1), \mathscr{K}_2, \ldots, \mathscr{K}_n) + \sum_{i=2}^{n} \mathscr{Q}_n(\mathscr{K}_1, \ldots, \mathscr{K}_{i-1}, \mathscr{D}(\mathscr{K}_i), \mathscr{K}_{i+1}, \ldots, \mathscr{K}_n) \tag{1.2}$$

holds for all $\mathscr{K}_1, \mathscr{K}_2, \ldots, \mathscr{K}_n \in \mathscr{A}$. Note that any nonlinear bi-skew Lie n-derivation is an example of a nonlinear generalized bi-skew Lie n-derivation (set $\mathscr{G} = \mathscr{D}$ in (1.2)).

Over the past decade, the structure of arbitrary maps involving various products on different kinds of $*$-algebras has been extensively studied by many authors. For examples, Li et al. [14] showed that every nonlinear skew Lie (resp. Jordan) derivation on a von Neumann algebra without central summands of type I_1 is an additive $*$-derivation. This result has been extended to the case of nonlinear $*$-Lie (Jordan)-type derivations by Lin in [11, 12]. Kong and Zhang [9] introduced the concept of the bi-skew Lie product and proved that a map $\mathscr{D}$ (not necessarily linear) on a factor von Neumann algebra $\mathscr{A}$ is a nonlinear bi-skew Lie 2-derivation if and only if $\mathscr{D}$ is an additive $*$-derivation. The authors [3] extended the above result to the case of nonlinear bi-skew Lie n-derivation on a factor von Neumann algebra. Recently, the authors in [2] introduced the bi-skew Jordan product ($\mathscr{K}_1 \diamond \mathscr{K}_2 = \mathscr{K}_1\mathscr{K}_2^* + \mathscr{K}_2\mathscr{K}_1^*$) and proved that each nonlinear bi-skew Jordan n-derivation on a factor von Neumann algebra is an additive $*$-derivation. However, the authors believe that it would be interesting to investigate these problems for arbitrary $*$-algebras. The present chapter deals with the afore-mentioned problem in the case of nonlinear generalized bi-skew Lie n-derivations of unital $*$-algebras. Precisely speaking, it demonstrates that under some mild restrictions, any nonlinear generalized bi-skew Lie n-derivation $\mathscr{G} : \mathscr{A} \to \mathscr{A}$ is of the form $\mathscr{G}(\mathscr{K}) = \mathscr{V}\mathscr{K} + \mathscr{D}(\mathscr{K})$ for all $\mathscr{K} \in \mathscr{A}$, where $\mathscr{V} \in \mathscr{Z}(\mathscr{A})$ (the center of $\mathscr{A}$) and $\mathscr{D} : \mathscr{A} \to \mathscr{A}$ is an additive $*$-derivation (Theorem 1.1). Finally, we provide a description of a nonlinear generalized bi-skew Lie n-derivation for some classical examples of unital $*$-algebras such as unital prime $*$-algebras (Corollary 1.1), standard operator algebras (Corollary 1.2), factor von Neumann algebras (Corollary 1.3), and von Neumann algebras with no central summands of type $\mathscr{I}_1$ (Corollary 1.4). Our study is primarily motivated by the content of the papers [1, 4]. Benkovic [4] recently described the form of generalized Lie n-derivations of a unital algebra $\mathscr{A}$ with a nontrivial idempotent and proved that under certain assumptions every generalized Lie n-derivation $\mathscr{G} : \mathscr{A} \to \mathscr{A}$ is of the form $\mathscr{G}(\mathscr{K}) = \mathscr{V}\mathscr{K} + \mathscr{D}(\mathscr{K})$

for all $\mathscr{K} \in \mathscr{A}$, where $\mathscr{V} \in \mathscr{Z}(\mathscr{A})$ and $\mathscr{D} : \mathscr{A} \to \mathscr{A}$ is a Lie n-derivation. In [1], Ashraf and Ansari extended the above result to the case of nonlinear generalized Lie n-derivations.

To set the stage, we introduce some notations and terminologies which will play a crucial role in establishing the validity of our main results. Let us assume that $\mathscr{A}$ is a unital $*$-algebra with the center $\mathscr{Z}(\mathscr{A})$, the identity element $\mathscr{I}$ and a nontrivial projection $\mathscr{P}$ (that is, $\mathscr{P} \neq 0, \mathscr{I}$ and $\mathscr{P}^2 = \mathscr{P} = \mathscr{P}^*$), and write $\mathscr{P}_1 = \mathscr{P}$ and $\mathscr{P}_2 = \mathscr{I} - \mathscr{P}_1$. Then $\mathscr{A}$ can be written as $\mathscr{A} = \mathscr{P}_1\mathscr{A}\mathscr{P}_1 + \mathscr{P}_1\mathscr{A}\mathscr{P}_2 + \mathscr{P}_2\mathscr{A}\mathscr{P}_1 + \mathscr{P}_2\mathscr{A}\mathscr{P}_2$. Let $\mathscr{A}^s = \{\mathscr{M} \in \mathscr{A} : \mathscr{M}^* = \mathscr{M}\}$ and $\mathscr{A}^k = \{\mathscr{N} \in \mathscr{A} : \mathscr{N}^* = -\mathscr{N}\}$. Furthermore, assume that $\mathscr{A}$ satisfies the following property

$$\mathscr{K}\mathscr{A}\mathscr{P}_i = \{0\} \text{ implies } \mathscr{K} = 0 \ (i = 1,2). \tag{1.3}$$

Some examples of unital $*$-algebras with nontrivial projections satisfying the property (1.3) are unital prime $*$-algebras, factor von Neumann algebras, von Neumann algebras with no central summands of type I_1 (see [6, 14] for details).

1.2 NONLINEAR GENERALIZED BI-SKEW LIE N-DERIVATIONS

This section is dedicated to unveiling the central outcome of this paper which in fact characterize nonlinear generalized bi-skew Lie n-derivations of unital $*$-algebras with nontrivial projections. Observe that if $\mathscr{G} : \mathscr{A} \to \mathscr{A}$ is a nonlinear generalized bi-skew Lie n-derivation with associated nonlinear bi-skew Lie n-derivation $\mathscr{D} : \mathscr{A} \to \mathscr{A}$, then it follows from (1.1) and (1.2) that a map $\mathscr{C} = \mathscr{G} - \mathscr{D}$ satisfies

$$\mathscr{C}(\mathscr{Q}_n(\mathscr{K}_1, \mathscr{K}_2, \ldots, \mathscr{K}_n)) = \mathscr{Q}_n(\mathscr{C}(\mathscr{K}_1), \mathscr{K}_2, \ldots, \mathscr{K}_n) \tag{1.4}$$

for all $\mathscr{K}_1, \mathscr{K}_2, \ldots, \mathscr{K}_n \in \mathscr{A}$. Therefore, to obtain a description of the map $\mathscr{G}$ it suffices to consider a map $\mathscr{C}$ with property (1.4). The following proposition gives a form of such maps on unital $*$-algebras with nontrivial projections.

Proposition 1.1 *Let $\mathscr{A}$ be a unital $*$-algebra with a nontrivial projection. If a map $\mathscr{C} : \mathscr{A} \to \mathscr{A}$ satisfies $\mathscr{C}(\mathscr{Q}_n(\mathscr{K}_1, \ldots, \mathscr{K}_n)) = \mathscr{Q}_n(\mathscr{C}(\mathscr{K}_1), \ldots, \mathscr{K}_n)$ for all $\mathscr{K}_1, \mathscr{K}_2, \ldots, \mathscr{K}_n \in \mathscr{A}$, then it is of the form $\mathscr{C}(\mathscr{K}) = \mathscr{V}\mathscr{K}$, where $\mathscr{V} \in \mathscr{Z}(\mathscr{A})$.*

The proof of the proposition can be achieved through the following series of lemmas.

Lemma 1.1 *For any $\mathscr{K}_1, \mathscr{K}_2, \ldots \mathscr{K}_n \in \mathscr{A}$,*

$$\mathscr{C}(\mathscr{Q}_n(\mathscr{K}_1, \mathscr{K}_2, \mathscr{K}_3, \ldots, \mathscr{K}_n)) = \mathscr{Q}_n(\mathscr{C}(\mathscr{K}_1), \mathscr{K}_2, \mathscr{K}_3, \ldots, \mathscr{K}_n)$$

if and only if

$$\mathscr{C}(\mathscr{Q}_n(\mathscr{K}_1, \mathscr{K}_2, \mathscr{K}_3, \ldots, \mathscr{K}_n)) = \mathscr{Q}_n(\mathscr{K}_1, \mathscr{C}(\mathscr{K}_2), \mathscr{K}_3, \ldots, \mathscr{K}_n).$$

Proof For 'if' part, assume that

$$\mathscr{C}(\mathscr{Q}_n(\mathscr{K}_1, \mathscr{K}_2, \mathscr{K}_3, \ldots, \mathscr{K}_n)) = \mathscr{Q}_n(\mathscr{C}(\mathscr{K}_1), \mathscr{K}_2, \mathscr{K}_3, \ldots, \mathscr{K}_n)$$

for all $\mathscr{K}_1, \ldots, \mathscr{K}_n \in \mathscr{A}$. Since $\mathscr{Q}_n(\mathscr{K}_1, \mathscr{K}_2, \mathscr{K}_3, \ldots, \mathscr{K}_n) = \mathscr{Q}_{n-1}([\mathscr{K}_1, \mathscr{K}_2]_\diamond, \mathscr{K}_3, \ldots, \mathscr{K}_n)$, we get

$$\mathscr{C}(\mathscr{Q}_{n-1}([\mathscr{K}_1, \mathscr{K}_2]_\diamond, \mathscr{K}_3, \ldots, \mathscr{K}_n)) = \mathscr{Q}_{n-1}([\mathscr{C}(\mathscr{K}_1), \mathscr{K}_2]_\diamond, \mathscr{K}_3, \ldots, \mathscr{K}_n).$$

Interchanging $\mathscr{K}_1$ and $\mathscr{K}_2$ in the above expression, we obtain

$$\mathscr{C}(\mathscr{Q}_{n-1}([\mathscr{K}_2, \mathscr{K}_1]_\diamond, \mathscr{K}_3, \ldots, \mathscr{K}_n)) = \mathscr{Q}_n([\mathscr{C}(\mathscr{K}_2), \mathscr{K}_1]_\diamond, \mathscr{K}_3, \ldots, \mathscr{K}_n).$$

Replacing $\mathscr{K}_1$ by $-\mathscr{K}_1$, we get

$$\mathscr{C}(\mathscr{Q}_{n-1}([\mathscr{K}_2,-\mathscr{K}_1]_\diamond,\mathscr{K}_3,\ldots,\mathscr{K}_n)) = \mathscr{Q}_n([\mathscr{C}(\mathscr{K}_2),-\mathscr{K}_1]_\diamond,\mathscr{K}_3,\ldots,\mathscr{K}_n).$$

Since $[\mathscr{K}_1,\mathscr{K}_2]_\diamond = -[\mathscr{K}_2,\mathscr{K}_1]_\diamond$ for all $\mathscr{K}_1,\mathscr{K}_2\in\mathscr{A}$, we have

$$\mathscr{C}(\mathscr{Q}_{n-1}([\mathscr{K}_1,\mathscr{K}_2]_\diamond,\mathscr{K}_3,\ldots,\mathscr{K}_n)) = \mathscr{Q}_{n-1}([\mathscr{K}_1,\mathscr{C}(\mathscr{K}_2)]_\diamond,\mathscr{K}_3,\ldots,\mathscr{K}_n).$$

This implies that

$$\mathscr{C}(\mathscr{Q}_n(\mathscr{K}_1,\mathscr{K}_2,\mathscr{K}_3,\ldots,\mathscr{K}_n)) = \mathscr{Q}_n(\mathscr{K}_1,\mathscr{C}(\mathscr{K}_2),\mathscr{K}_3,\ldots,\mathscr{K}_n)$$

for all $\mathscr{K}_1,\ldots,\mathscr{K}_n\in\mathscr{A}$. Similarly, one can verify the 'only if' part.

Lemma 1.2 *$\mathscr{C}$ is additive.*

Proof For any $\mathscr{K}_1,\mathscr{K}_2\in\mathscr{A}$, set $\mathscr{S}=\mathscr{C}(\mathscr{K}_1+\mathscr{K}_2)-\mathscr{C}(\mathscr{K}_1)-\mathscr{C}(\mathscr{K}_2)$. Our aim is to show $\mathscr{S}=0$. For any $\mathscr{V}\in\mathscr{A}$, using Lemma 1.1, we get

$$\begin{aligned}\mathscr{Q}_n(\mathscr{C}(\mathscr{K}_1+\mathscr{K}_2),\mathscr{V},\mathscr{I},\ldots,\mathscr{I}) &= \mathscr{Q}_n(\mathscr{K}_1+\mathscr{K}_2,\mathscr{C}(\mathscr{V}),\mathscr{I},\ldots,\mathscr{I})\\ &= \mathscr{Q}_n(\mathscr{K}_1,\mathscr{C}(\mathscr{V}),\mathscr{I},\ldots,\mathscr{I})+\mathscr{Q}_n(\mathscr{K}_2,\mathscr{C}(\mathscr{V}),\mathscr{I},\ldots,\mathscr{I})\\ &= \mathscr{Q}_n(\mathscr{C}(\mathscr{K}_1),\mathscr{V},\mathscr{I},\ldots,\mathscr{I})+\mathscr{Q}_n(\mathscr{C}(\mathscr{K}_2),\mathscr{V},\mathscr{I},\ldots,\mathscr{I})\\ &= \mathscr{Q}_n(\mathscr{C}(\mathscr{K}_1)+\mathscr{C}(\mathscr{K}_2),\mathscr{V},\mathscr{I},\ldots,\mathscr{I}).\end{aligned}$$

This implies that

$$\mathscr{Q}_n(\mathscr{S},\mathscr{V},\mathscr{I},\ldots,\mathscr{I})=0 \text{ for all } \mathscr{V}\in\mathscr{A}. \tag{1.5}$$

Choosing $\mathscr{V}=\mathscr{I}$ in (1.5) and simplifying, we get $\mathscr{S}^*=\mathscr{S}$. Similarly, taking $\mathscr{V}=\mathrm{i}\mathscr{I}$ in (1.5) and simplifying, we obtain $\mathscr{S}^*=-\mathscr{S}$. Thus, $\mathscr{S}=0$, and hence $\mathscr{D}$ is additive.

Lemma 1.3 *For any $\mathscr{K}\in\mathscr{A}$, we have $\mathscr{C}(\mathrm{i}\mathscr{K})=i\mathscr{C}(\mathscr{K})$ and $\mathscr{C}(\mathscr{K})^*=\mathscr{C}(\mathscr{K})$.*

Proof First, we show that $\mathscr{C}(\mathscr{M})^*=\mathscr{C}(\mathscr{M})$ and $\mathscr{C}(\mathrm{i}\mathscr{M})=\mathrm{i}\mathscr{C}(\mathscr{M})$ for all $\mathscr{M}\in\mathscr{A}^s$. Let $\mathscr{M}\in\mathscr{A}^s$. Using the fact that $\mathscr{Q}_n\left(\mathscr{M},\frac{\mathscr{I}}{2},\ldots,\frac{\mathscr{I}}{2}\right)=0$, we have

$$0=\mathscr{C}\left(\mathscr{Q}_n\left(\mathscr{M},\frac{\mathscr{I}}{2},\ldots,\frac{\mathscr{I}}{2}\right)\right)=\mathscr{Q}_n\left(\mathscr{C}(\mathscr{M}),\frac{\mathscr{I}}{2},\ldots,\frac{\mathscr{I}}{2}\right)=\left(\frac{\mathscr{I}}{2}\right)\{\mathscr{C}(\mathscr{M})-\mathscr{C}(\mathscr{M})^*\},$$

which in turn implies that $\mathscr{C}(\mathscr{M})^*=\mathscr{C}(\mathscr{M})$.

For any $\mathscr{M}\in\mathscr{A}^s$, we have $\mathscr{Q}_n(\mathscr{M},\mathrm{i}\mathscr{I},\mathscr{I},\ldots,\mathscr{I})=-2^{n-1}\mathrm{i}\mathscr{M}$. Using Lemma 1.2 and the fact that $\mathscr{C}(\mathscr{M})^*=\mathscr{C}(\mathscr{M})$, we get

$$\begin{aligned}-2^{n-1}\mathscr{C}(\mathrm{i}\mathscr{M})=\mathscr{C}(-2^{n-1}\mathrm{i}\mathscr{M}) &= \mathscr{C}(\mathscr{Q}_n(\mathscr{M},\mathrm{i}\mathscr{I},\mathscr{I},\ldots,\mathscr{I})),\\ &= \mathscr{Q}_n(\mathscr{C}(\mathscr{M}),\mathrm{i}\mathscr{I},\mathscr{I},\ldots,\mathscr{I}),\\ &= -2^{n-1}\mathrm{i}\mathscr{C}(\mathscr{M}),\end{aligned}$$

which gives $\mathscr{C}(\mathrm{i}\mathscr{M})=\mathrm{i}\mathscr{C}(\mathscr{M})$.

Observe that any $\mathscr{K}\in\mathscr{A}$ can be written as $\mathscr{K}=\frac{\mathscr{K}+\mathscr{K}^*}{2}+\mathrm{i}\frac{\mathscr{K}-\mathscr{K}^*}{2\mathrm{i}}$, where $\frac{\mathscr{K}+\mathscr{K}^*}{2},\frac{\mathscr{K}-\mathscr{K}^*}{2\mathrm{i}}\in\mathscr{A}^s$. Hence, without loss of generality, we assume that $\mathscr{K}=\mathscr{M}_1+\mathrm{i}\mathscr{M}_2$, where $\mathscr{M}_1,\mathscr{M}_2\in\mathscr{A}^s$. Using Lemma 1.2, we obtain

$$\begin{aligned}\mathscr{C}(\mathrm{i}\mathscr{K}) &= \mathscr{C}(\mathrm{i}\mathscr{M}_1-\mathscr{M}_2)\\ &= \mathscr{C}(\mathrm{i}\mathscr{M}_1)-\mathscr{C}(\mathscr{M}_2)\\ &= \mathrm{i}\mathscr{C}(\mathscr{M}_1)-\mathscr{C}(\mathscr{M}_2)\\ &= \mathrm{i}\{\mathscr{C}(\mathscr{M}_1)+\mathrm{i}\mathscr{C}(\mathscr{M}_2)\}\\ &= \mathrm{i}\mathscr{C}(\mathscr{M}_1+\mathrm{i}\mathscr{M}_2)\\ &= \mathrm{i}\mathscr{C}(\mathscr{K}).\end{aligned}$$

Next, using Lemma 1.2, we obtain

$$\begin{aligned}\mathscr{C}(\mathscr{K})^* &= \mathscr{C}(\mathscr{M}_1+\mathrm{i}\mathscr{M}_2)^* = \mathscr{C}(\mathscr{M}_1)^* - \mathrm{i}\mathscr{C}(\mathscr{M}_2)^* = \mathscr{C}(\mathscr{M}_1^*) - \mathrm{i}\mathscr{C}(\mathscr{M}_2^*) = \mathscr{C}(\mathscr{M}_1^* - \mathrm{i}\mathscr{M}_2^*)\\ &= \mathscr{C}(\mathscr{K}^*).\end{aligned}$$

Lemma 1.4 *$\mathscr{C}(\mathscr{K}_1\mathscr{K}_2) = \mathscr{C}(\mathscr{K}_1)\mathscr{K}_2 = \mathscr{K}_1\mathscr{C}(\mathscr{K}_2)$ for all $\mathscr{K}_1, \mathscr{K}_2 \in \mathscr{A}$.*

Proof First, we show that $\mathscr{C}(\mathscr{M}_1\mathscr{M}_2) = \mathscr{C}(\mathscr{M}_1)\mathscr{M}_2 = \mathscr{M}_1\mathscr{C}(\mathscr{M}_2)$ for all $\mathscr{M}_1, \mathscr{M}_2 \in \mathscr{A}^s$. For any $\mathscr{M}_1, \mathscr{M}_2 \in \mathscr{A}^s$, we have $\mathscr{Q}_n(\mathscr{M}_1, \mathscr{M}_2, \mathscr{I}, \ldots, \mathscr{I}) = 2^{n-2}(\mathscr{M}_1\mathscr{M}_2 - \mathscr{M}_2\mathscr{M}_1)$. Using Lemmas 1.2 and 1.3, we obtain

$$\begin{aligned}2^{n-2}\mathscr{C}(\mathscr{M}_1\mathscr{M}_2 - \mathscr{M}_2\mathscr{M}_1) &= \mathscr{C}(\mathscr{Q}_n(\mathscr{M}_1, \mathscr{M}_2, \mathscr{I}, \ldots, \mathscr{I}))\\ &= \mathscr{Q}_n(\mathscr{C}(\mathscr{M}_1), \mathscr{M}_2, \mathscr{I}, \ldots, \mathscr{I})\\ &= \mathscr{Q}_{n-1}(\mathscr{C}(\mathscr{M}_1)\mathscr{M}_2 - \mathscr{M}_2\mathscr{C}(\mathscr{M}_1), \mathscr{I}, \ldots, \mathscr{I})\\ &= 2^{n-2}\{\mathscr{C}(\mathscr{M}_1)\mathscr{M}_2 - \mathscr{M}_2\mathscr{C}(\mathscr{M}_1)\}.\end{aligned}$$

That is,

$$\mathscr{C}(\mathscr{M}_1\mathscr{M}_2) - \mathscr{C}(\mathscr{M}_2\mathscr{M}_1) = \mathscr{C}(\mathscr{M}_1)\mathscr{M}_2 - \mathscr{M}_2\mathscr{C}(\mathscr{M}_1). \tag{1.6}$$

Also, $\mathscr{Q}_n(\mathrm{i}\mathscr{M}_1, \mathscr{M}_2, \mathscr{I}, \ldots, \mathscr{I}) = 2^{n-2}\mathrm{i}\{\mathscr{M}_1\mathscr{M}_2 + \mathscr{M}_2\mathscr{M}_1\}$. Using Lemmas 1.2 and 1.3, we find that

$$\begin{aligned}2^{n-2}\mathrm{i}\mathscr{C}(\mathscr{M}_1\mathscr{M}_2 + \mathscr{M}_2\mathscr{M}_1) &= \mathscr{C}(2^{n-2}\mathrm{i}(\mathscr{M}_1\mathscr{M}_2 + \mathscr{M}_2\mathscr{M}_1))\\ &= \mathscr{C}(\mathscr{Q}_n(\mathrm{i}\mathscr{M}_1, \mathscr{M}_2, \mathscr{I}, \ldots, \mathscr{I}))\\ &= \mathscr{Q}_n(\mathscr{C}(\mathrm{i}\mathscr{M}_1), \mathscr{M}_2, \mathscr{I}, \ldots, \mathscr{I})\\ &= \mathscr{Q}_n(\mathrm{i}\mathscr{C}(\mathscr{M}_1), \mathscr{M}_2, \mathscr{I}, \ldots, \mathscr{I})\\ &= \mathscr{Q}_{n-1}(\mathrm{i}\mathscr{C}(\mathscr{M}_1)\mathscr{M}_2 + \mathrm{i}\mathscr{M}_2\mathscr{C}(\mathscr{M}_1), \mathscr{I}, \ldots, \mathscr{I})\\ &= 2^{n-2}\mathrm{i}\{\mathscr{C}(\mathscr{M}_1)\mathscr{M}_2 + \mathscr{M}_2\mathscr{C}(\mathscr{M}_1)\}.\end{aligned}$$

That is,

$$\mathscr{C}(\mathscr{M}_1\mathscr{M}_2) + \mathscr{C}(\mathscr{M}_2\mathscr{M}_1) = \mathscr{C}(\mathscr{M}_1)\mathscr{M}_2 + \mathscr{M}_2\mathscr{C}(\mathscr{M}_1). \tag{1.7}$$

Adding the expressions (1.6) and (1.7), we get $\mathscr{C}(\mathscr{M}_1\mathscr{M}_2) = \mathscr{C}(\mathscr{M}_1)\mathscr{M}_2$. Further, subtraction of (1.6) from (1.7) yields that $\mathscr{C}(\mathscr{M}_2\mathscr{M}_1) = \mathscr{M}_2\mathscr{C}(\mathscr{M}_1)$. Therefore,

$$\mathscr{C}(\mathscr{M}_1\mathscr{M}_2) = \mathscr{C}(\mathscr{M}_1)\mathscr{M}_2 = \mathscr{M}_2\mathscr{C}(\mathscr{M}_1) \text{ for all } \mathscr{M}_1, \mathscr{M}_2 \in \mathscr{A}^s.$$

Now, take $\mathscr{K}_1 = \mathscr{M}_1 + \mathrm{i}\mathscr{M}_2, \mathscr{K}_2 = \mathscr{M}_1' + \mathrm{i}\mathscr{M}_2' \in \mathscr{A}$, where $\mathscr{M}_1, \mathscr{M}_1', \mathscr{M}_2, \mathscr{M}_2' \in \mathscr{A}^s$. Using Lemmas 1.2 and 1.3, we get

$$\begin{aligned}\mathscr{C}(\mathscr{K}_1\mathscr{K}_2) &= \mathscr{C}((\mathscr{M}_1\mathscr{M}_1' - \mathscr{M}_2\mathscr{M}_2') + \mathrm{i}(\mathscr{M}_1\mathscr{M}_2' + \mathscr{M}_2\mathscr{M}_1'))\\ &= \mathscr{C}(\mathscr{M}_1\mathscr{M}_1') - \mathscr{C}(\mathscr{M}_2\mathscr{M}_2') + \mathrm{i}\mathscr{C}(\mathscr{M}_1\mathscr{M}_2') + \mathrm{i}\mathscr{C}(\mathscr{M}_2\mathscr{M}_1')\\ &= \mathscr{C}(\mathscr{M}_1)\mathscr{M}_1' - \mathscr{C}(\mathscr{M}_2)\mathscr{M}_2' + \mathrm{i}\mathscr{C}(\mathscr{M}_1)\mathscr{M}_2' + \mathrm{i}\mathscr{C}(\mathscr{M}_2)\mathscr{M}_1'.\end{aligned}$$

Also,

$$\begin{aligned}\mathscr{C}(\mathscr{K}_1)\mathscr{K}_2 &= \mathscr{C}(\mathscr{M}_1 + \mathrm{i}\mathscr{M}_2)(\mathscr{M}_1' + \mathrm{i}\mathscr{M}_2')\\ &= (\mathscr{C}(\mathscr{M}_1) + \mathrm{i}\mathscr{C}(\mathscr{M}_2))(\mathscr{M}_1' + \mathrm{i}\mathscr{M}_2')\\ &= \mathscr{C}(\mathscr{M}_1)\mathscr{M}_1' - \mathscr{C}(\mathscr{M}_2)\mathscr{M}_2' + \mathrm{i}\mathscr{C}(\mathscr{M}_1)\mathscr{M}_2' + \mathrm{i}\mathscr{C}(\mathscr{M}_2)\mathscr{M}_1'.\end{aligned}$$

Thus, $\mathscr{C}(\mathscr{K}_1\mathscr{K}_2) = \mathscr{C}(\mathscr{K}_1)\mathscr{K}_2$ for all $\mathscr{K}_1, \mathscr{K}_2 \in \mathscr{A}$. Similarly, one can obtain that $\mathscr{C}(\mathscr{K}_1\mathscr{K}_2) = \mathscr{K}_1\mathscr{C}(\mathscr{K}_2)$ for all $\mathscr{K}_1, \mathscr{K}_2 \in \mathscr{A}$.

Lemma 1.5 *$\mathscr{C}(\mathscr{K}) = \mathscr{V}\mathscr{K}$ for all $\mathscr{K} \in \mathscr{A}$, where $\mathscr{V} \in \mathscr{Z}(\mathscr{A})$.*

Proof Let $\mathscr{K}$ be an arbitrary element in $\mathscr{A}$. Using Lemma 1.4, we have

$$\mathscr{C}(\mathscr{K}) = \mathscr{C}(\mathscr{I}\mathscr{K}) = \mathscr{K}\mathscr{C}(\mathscr{I}) \text{ and } \mathscr{C}(\mathscr{K}) = \mathscr{C}(\mathscr{K}\mathscr{I}) = \mathscr{C}(\mathscr{I})\mathscr{K}.$$

This implies that $\mathscr{C}(\mathscr{I}) \in \mathscr{Z}(\mathscr{A})$. Set $\mathscr{V} = \mathscr{C}(\mathscr{I})$. Then, it follows that $\mathscr{C}(\mathscr{K}) = \mathscr{V}\mathscr{K}$ for all $\mathscr{K} \in \mathscr{A}$, where $\mathscr{V} \in \mathscr{Z}(\mathscr{A})$ with $\mathscr{V}^* = \mathscr{V}$.

The following proposition provides the structure of nonlinear generalized bi-skew Lie n-derivations on unital $*$-algebras with nontrivial projections.

Proposition 1.2 *Let $\mathscr{A}$ be a unital $*$-algebra with a nontrivial projection. Then every nonlinear generalized bi-skew Lie n-derivation $\mathscr{G} : \mathscr{A} \to \mathscr{A}$ with associated nonlinear bi-skew Lie n-derivation $\mathscr{D} : \mathscr{A} \to \mathscr{A}$ is of the form $\mathscr{G}(\mathscr{K}) = \mathscr{V}\mathscr{K} + \mathscr{D}(\mathscr{K})$ for all $\mathscr{K} \in \mathscr{A}$, where $\mathscr{V} \in \mathscr{Z}(\mathscr{A})$.*

Proof Let $\mathscr{G} : \mathscr{A} \to \mathscr{A}$ be a nonlinear generalized bi-skew Lie n-derivation with associated nonlinear bi-skew Lie n-derivation $\mathscr{D} : \mathscr{A} \to \mathscr{A}$. Then, the map $\mathscr{C} = \mathscr{G} - \mathscr{D}$ satisfies

$$\mathscr{C}(\mathscr{Q}_n(\mathscr{K}_1, \mathscr{K}_2, \ldots, \mathscr{K}_n)) = \mathscr{Q}_n(\mathscr{C}(\mathscr{K}_1), \mathscr{K}_2, \ldots, \mathscr{K}_n)$$

for all $\mathscr{K}_1, \mathscr{K}_2, \ldots, \mathscr{K}_n \in \mathscr{A}$. Hence, it follows from Proposition 12.1 that $\mathscr{C}(\mathscr{K}) = \mathscr{V}\mathscr{K}$ for all $\mathscr{K} \in \mathscr{A}$, where $\mathscr{V} \in \mathscr{Z}(\mathscr{A})$. Consequently, $\mathscr{G}(\mathscr{K}) = \mathscr{V}\mathscr{K} + \mathscr{D}(\mathscr{K})$ for all $\mathscr{K} \in \mathscr{A}$, where $\mathscr{V} \in \mathscr{Z}(\mathscr{A})$. This completes the proof of the proposition.

Let $\mathscr{A}$ be a unital $*$-algebra with a nontrivial idempotent $\mathscr{P}_1$ satisfying (1.3). It is shown in [26, Theorem 2.1] that every nonlinear bi-skew Lie n-derivation $\mathscr{D} : \mathscr{A} \to \mathscr{A}$ is an additive $*$-derivation. Combining this result with Proposition 1.2, we obtain the main result of the chapter.

Theorem 1.1 *Let $\mathscr{A}$ be a unital $*$-algebra with a nontrivial projection $\mathscr{P}_1$ satisfying* (1.3). *Then every nonlinear generalized bi-skew Lie n-derivation $\mathscr{G} : \mathscr{A} \to \mathscr{A}$ is of the form $\mathscr{G}(\mathscr{K}) = \mathscr{V}\mathscr{K} + \mathscr{D}(\mathscr{K})$ for all $\mathscr{K} \in \mathscr{A}$, where $\mathscr{V} \in \mathscr{Z}(\mathscr{A})$ and $\mathscr{D} : \mathscr{A} \to \mathscr{A}$ is an additive $*$-derivation. Moreover, $\mathscr{G} : \mathscr{A} \to \mathscr{A}$ is an additive generalized $*$-derivation.*

Proof Let $\mathscr{G} : \mathscr{A} \to \mathscr{A}$ be a nonlinear generalized bi-skew Lie n-derivation with associated nonlinear bi-skew Lie n-derivation $\mathscr{D} : \mathscr{A} \to \mathscr{A}$. Then, it follow from Proposition 1.2 that $\mathscr{G}$ is of the form $\mathscr{G}(\mathscr{K}) = \mathscr{V}\mathscr{K} + \mathscr{D}(\mathscr{K})$ for all $\mathscr{K} \in \mathscr{A}$, where $\mathscr{V} \in \mathscr{Z}(\mathscr{A})$. Since $\mathscr{D} : \mathscr{A} \to \mathscr{A}$ is a nonlinear bi-skew Lie n-derivation, in view of [26, Theorem 2.1], $\mathscr{D}$ is an additive $*$-derivation. It only remains to show that $\mathscr{G}$ is an additive generalized $*$-derivation. We assert that $\mathscr{G}$ is an additive generalized $*$-derivation with associated additive $*$-derivation $\mathscr{D}$. Clearly, $\mathscr{G}$ is additive. For any $\mathscr{K}_1, \mathscr{K}_2 \in \mathscr{A}$, we have

$$\begin{aligned}
\mathscr{G}(\mathscr{K}_1)\mathscr{K}_2 + \mathscr{K}_1\mathscr{D}(\mathscr{K}_2) &= (\mathscr{V}\mathscr{K}_1 + \mathscr{D}(\mathscr{K}_1))\mathscr{K}_2 + \mathscr{K}_1\mathscr{D}(\mathscr{K}_2) \\
&= (\mathscr{V}\mathscr{K}_1)\mathscr{K}_2 + \mathscr{D}(\mathscr{K}_1)\mathscr{K}_2 + \mathscr{K}_1\mathscr{D}(\mathscr{K}_2) \\
&= \mathscr{V}(\mathscr{K}_1\mathscr{K}_2) + \mathscr{D}(\mathscr{K}_1\mathscr{K}_2) \\
&= \mathscr{G}(\mathscr{K}_1\mathscr{K}_2).
\end{aligned}$$

We now show that $\mathscr{G}(\mathscr{K}^*) = \mathscr{G}(\mathscr{K})^*$ for all $\mathscr{K} \in \mathscr{A}$. Define a map $\mathscr{C} = \mathscr{G} - \mathscr{D}$. Then, $\mathscr{C}$ satisfies (1.4) and hence, by Lemma 1.3, $\mathscr{C}(\mathscr{K}^*) = \mathscr{C}(\mathscr{K})^*$ for all $\mathscr{K} \in \mathscr{A}$. Thus, for any $\mathscr{K} \in \mathscr{A}$, we get

$$\mathscr{G}(\mathscr{K}^*) = \mathscr{C}(\mathscr{K}^*) - \mathscr{D}(\mathscr{K}^*) = \mathscr{C}(\mathscr{K})^* - \mathscr{D}(\mathscr{K})^* = (\mathscr{C}(\mathscr{K}) - \mathscr{D}(\mathscr{K}))^* = \mathscr{G}(\mathscr{K})^*.$$

Consequently, $\mathscr{G}$ is an additive generalized $*$-derivation with associated additive $*$-derivation $\mathscr{D}$.

1.3 APPLICATIONS

In this section, we put into practice Theorem 1.1 to certain special classes of unital $*$-algebras such as prime $*$-algebras, standard operator algebras, factor von Neumann algebras, and von Neumann algebras with no central summands of type I_1.

Recall that an algebra $\mathscr{A}$ is *prime* if for any $\mathscr{K}_1, \mathscr{K}_2 \in \mathscr{A}$, the condition $\mathscr{K}_1\mathscr{A}\mathscr{K}_2 = \{0\}$ leads to the conclusion that either $\mathscr{K}_1 = 0$ or $\mathscr{K}_2 = 0$. It is easy to see that every prime $*$-algebra with a nontrivial projection satisfies the property (1.3). Hence, in view of Theorem 1.1, we arrive at the following result:

Corollary 1.1 *Let $\mathscr{A}$ be a unital prime $*$-algebra with a nontrivial projection. Then every nonlinear generalized bi-skew Lie n-derivation $\mathscr{G} : \mathscr{A} \to \mathscr{A}$ is of the form $\mathscr{G}(\mathscr{K}) = \mathscr{V}\mathscr{K} + \mathscr{D}(\mathscr{K})$ for all $\mathscr{K} \in \mathscr{A}$, where $\mathscr{V} \in \mathscr{Z}(\mathscr{A})$ and $\mathscr{D} : \mathscr{A} \to \mathscr{A}$ is an additive $*$-derivation. Moreover, $\mathscr{G} : \mathscr{A} \to \mathscr{A}$ is an additive generalized $*$-derivation.*

Let $\mathscr{H}$ be a complex Hilbert space, and $\mathscr{B}(\mathscr{H})$ be the algebra of all bounded linear operators on $\mathscr{H}$. Let $\mathscr{F}(\mathscr{H}) \subseteq \mathscr{B}(\mathscr{H})$ denote the subalgebra of all bounded finite rank operators. A subalgebra $\mathscr{A} \subseteq \mathscr{B}(\mathscr{H})$ is called a *standard operator algebra* if it contains $\mathscr{F}(\mathscr{H})$ and the identity operator. If $\mathscr{A}$ is closed under the adjoint operation, then it becomes a unital prime $*$-algebra with a nontrivial projection. Hence, in view of Corollary 1.1, we have the following result:

Corollary 1.2 *Let $\mathscr{H}$ be an infinite-dimensional complex Hilbert space and $\mathscr{A} \subseteq \mathscr{B}(\mathscr{H})$ be a standard operator algebra. Suppose that $\mathscr{A}$ is closed under the adjoint operation. Then every nonlinear generalized bi-skew Lie n-derivation $\mathscr{G} : \mathscr{A} \to \mathscr{A}$ is of the form $\mathscr{G}(\mathscr{K}) = \mathscr{V}\mathscr{K} + \mathscr{D}(\mathscr{K})$ for all $\mathscr{K} \in \mathscr{A}$, where $\mathscr{V} \in \mathscr{Z}(\mathscr{A})$ and $\mathscr{D} : \mathscr{A} \to \mathscr{A}$ is an additive $*$-derivation. Moreover, $\mathscr{G} : \mathscr{A} \to \mathscr{A}$ is an additive generalized $*$-derivation.*

A subalgebra $\mathscr{A} \subseteq \mathscr{B}(\mathscr{H})$ is called a *von Neumann algebra* if it is both weakly closed, self-adjoint, and contains the identity operator $\mathscr{I}$. Furthermore, a von Neumann algebra is called factor if its *center* is $\mathbb{C}\mathscr{I}$. Since a factor von Neumann algebra is prime, with direct reference to Corollary 1.1, we get:

Corollary 1.3 *Let $\mathscr{A}$ be a factor von Neumann algebra with dim$(\mathscr{A}) \geq 2$. Then every nonlinear generalized bi-skew Lie n-derivation $\mathscr{G} : \mathscr{A} \to \mathscr{A}$ is of the form $\mathscr{G}(\mathscr{K}) = \lambda\mathscr{K} + \mathscr{D}(\mathscr{K})$ for all $\mathscr{K} \in \mathscr{A}$, where $\lambda \in \mathbb{C}$ and $\mathscr{D} : \mathscr{A} \to \mathscr{A}$ is an additive $*$-derivation. Moreover, $\mathscr{G} : \mathscr{A} \to \mathscr{A}$ is an additive generalized $*$-derivation.*

Further, it is well-known that every von Neumann algebra having no central summands of type I_1 satisfies (1.3) (see [6, 14] for details). Thus, as a direct consequence of Theorem 1.1, we obtain the following corollary:

Corollary 1.4 *Let $\mathscr{A}$ be a von Neumann algebra having no central summands of type I_1. Then every nonlinear generalized bi-skew Lie n-derivation $\mathscr{G} : \mathscr{A} \to \mathscr{A}$ is of the form $\mathscr{G}(\mathscr{K}) = \mathscr{V}\mathscr{K} + \mathscr{D}(\mathscr{K})$ for all $\mathscr{K} \in \mathscr{A}$, where $\mathscr{V} \in \mathscr{Z}(\mathscr{A})$ and $\mathscr{D} : \mathscr{A} \to \mathscr{A}$ is an additive $*$-derivation. Moreover, $\mathscr{G} : \mathscr{A} \to \mathscr{A}$ is an additive generalized $*$-derivation.*

Acknowledgment: The authors would like to express their sincere thanks to the referee for careful reading of the chapter. This research is partially supported by a research grant from the National Board for Higher Mathematics, India (No. 02011/5/2020 NBHM(R.P.) R &D II/6243).

REFERENCES

1. Ashraf, M. & Ansari, M. A. (2022). Multiplicative generalized Lie n-derivations of unital rings with idempotents. *Rev. Real Acad. Cienc. Exactas Fis. Nat. Ser. A-Mat., 116(92).* https://doi.org/10.1007/s13398-022-01233-5
2. Ashraf, M., Akhter, M. S. & Ansari M. A. (2022). Nonlinear bi-skew Jordan-type derivations on factor von Neumann algebras. *Filomat, 37(17),* 5591–5599.
3. Ashraf, M., Akhter, M. S. & Ansari M. A. (2022). Nonlinear bi-skew Lie-type derivations on von Neumann algberas. *Comm. Algebra, 50(11),* 4766–4780.
4. Benkovic, D. (2018). Generalized Lie derivations of unital algebras with idempotents. *Oper. Matrices, 12(2),* 357–367.
5. Bresar, M. & Fosner, M. (2000). On rings with involution equipped with some new product. *Publ. Math. Debrecen, 57(1–2),* 121–134.
6. Dai, L. & Lu, F. (2014). Nonlinear maps preserving Jordan $*$-products. *J. Math. Anal. Appl., 409,* 180–188.
7. Ferreira, B. L. M. & Costa, B. T. (2021). $*$-Jordan-type maps on C^*-algebras. *Comm. Algebra, 49(12),* 5073–5082.
8. Ferreira, B. L. M. & Costa, B. T. (2022). $*$-Lie-Jordan-type maps on C^*-algebras. *Bull. Iran. Math. Soc., 48,* 1679–1690.
9. Kong, L. & Zhang, J. (2021). Nonlinear bi-skew Lie derivation on factor von Neumann algebras. *Bull. Iranian Math. Soc., 47(4),* 1097–1106.
10. Khan, A. N. (2021). Multiplicative bi-skew Lie triple derivations on factor von Neumann algebras. *Rocky Mountain J. Math., 51(6),* 2103–2114.
11. Lin, W.-H. (2018). Nonlinear $*$-Lie-type derivations on von Neumann algebras. *Acta Math. Hungar., 156(1),* 112–131.
12. Lin, W.-H. (2018). Nonlinear $*$-Jordan-type derivations on von Neumann algebras. https://arxiv.org/submit/2248854 arXiv: 1805.16027v1 [math.OA].
13. Li, C., Zhao, Y. & Zhao, F. (2021). Nonlinear $*$-Jordan-type derivations on $*$-algebras. *Rocky Mountain J. Math., 51(2),* 601–612.
14. Li, C., Lu, F. & Fang, X. (2014). Nonlinear ξ-Jordan $*$-derivations on von Neumann algebras. *Linear Multilinear Algebra 62,* 466–473.
15. Li, C.-J., Zhao, F.-F. & Chen, Q.-Y. (2016). Nonlinear skew Lie triple derivations between factors. *Acta Math. Sin. (Engl. Ser.)* **32***(7),* 821–830.
16. Molnar, L. (1997). Jordan $*$-derivation pairs on a complex $*$-algebra. *Aequationes Math., 54,* 44–55.
17. Molnar, L. & Semrl, P. (1997). Local Jordan $*$-derivations of standard operator algebras. *Proc. Amer. Math. Soc. 125,* 447–454.
18. Semrl, P. (1990). On Jordan $*$-derivations and an application. *Colloq. Math., 59,* 241–251.
19. Semrl, P. (1991). Quadratic functionals and Jordan $*$-derivations. *Studia Math., 97,* 157–165.
20. Semrl, P. (1994). Jordan $*$-derivations of standard operator algebras. *Proc. Amer. Math. Soc., 120(2),* 515–519.
21. Taghavi, A., Rohi, H. & Darvish, V. (2016). Nonlinear $*$-Jordan derivable maps on factor von Neumann algebras. *Linear Multilinear Algebra, 64(3),* 426–439.
22. Wani, B. A., Ashraf, M. & Lin, W.-H. (2020). Multiplicative $*$-Jordan type higher derivations on von Neumann algebras. *Quaest. Math., 43(12),* 1689–1711.
23. Wani, B. A., Ashraf, M. & Akhtar, M. S. (2021). Multiplicative $*$-Lie type higher derivations of standard operator algebras. *Comm. Algebra., 49(9),* 3777–3797.
24. Yu, W.-Y. & Zhang, J.-H. (2012). Nonlinear $*$-Lie derivations on factor von Neumann algebras. *Linear Algebra Appl., 437(8),* 1979–1991.
25. Zhang, F. (2016). Nonlinear skew Jordan derivable maps on factor von neumann algebras. *Linear Multilinear Algebra, 64(10),* 2090–2103.
26. Zhang, J. & Li, C.-J. (2023). Nonlinear bi-skew Lie-type derivations on $*$-algebras. *Rocky Mountain J. Math.* (to appear).

2 Love Wave Propagation in Fibre-Reinforced Layer Imperfectly Bonded to Micropolar Elastic Half-Space

Dr. Tanupreet Kaur
Department of Mathematics,
Guru Nanak College, Budhlada (Mansa), Punjab, India

CONTENTS

In nature, there exist some continuum whose microstructure plays very important role and it cannot be disregarded during investigation of problems related to wave propagation. Also, the bonding between the layer and half-space in the composite structure is not always perfect. Therefore, the present study investigates the dispersion of Love type wave in a vertically homogeneous fibre-reinforced layer lying over a micropolar elastic half-space. An imperfect interfacial bonding between layer and half-space is proposed. An analytically expression of dispersion equation of Love type wave has been derived analytically. The study reveals that wave number (k), coupling factor (N), and imperfectness parameter (Γ) disfavor the phase velocity of the Love wave, whereas reinforcement (μ_L/μ_T) and micropolarity (kH^2/γ) favor the phase velocity of the Love wave. The effect of complex interface on phase velocity is analysed as a particular case of the problem.

2.1 INTRODUCTION

In the layered medium problems, it has been generally assumed that there is no slipping at the interface between the layer and half-space. In such cases, the phenomena of imperfect bonding play

DOI: 10.1201/9781032634142-2

a crucial role in the functionality and reliability of materials. It is evident that imperfectly bonded materials are characterized by weaker effective moduli than perfectly bonded ones. However, in the most practical cases, the weak interface response leads to the development of dislocations and voids, which dramatically decreases the strength of the whole material. Therefore, there have been considerable interests in imperfect interface problems as may be appropriate in the case of either pre-existing defects or interface damage, for example, the cyclic loading. Imperfect bonding often occurs in surface acoustic wave (SAW) devices because of the aging of glue applied to two conjunct solids, micro defects, diffusion impurities, and other forms of damages. Hence, two dissimilar materials cannot be perfectly bonded, and an interphase or transition with a thickness typically within the range of 30−240 nm exists across the interface. Significant work has been done to describe the physical conditions on the interface by different mechanical boundary conditions by different investigators. Notable among them are Jones and Whitter [1], Murty [2], Nayfeh and Nassar [3], Lavrentyev and Rokhlin [4], Fan and Sze [5], Wang and Zhong [6], Shodja et al. [7], Shariat and Eslami [8]. Recently, Feng and Peng [9] studied the propagation of shear horizontal waves in a piezoelectric layer imperfectly bonded to a metal or elastic substrate. Imperfect bonding considered in present study means that the components of stress are continuous and the small displacement field is not continuous. The small vector difference in the displacement is assumed to depend linearly on the traction vector. The state of boundary significantly affects elastic wave propagation phenomena.

Due to low weight and high strength, fibre-reinforced composites are used in a variety of structures therefore, the study of their mechanical behavior has great importance in geomechanics, civil engineering, mining engineering, architecture, earthquake engineering, engineering geology, and geophysics. The earth medium has various types of hard and/or soft rocks with different elastic properties may be fibre-reinforcement in nature. The characteristic property of fibre-reinforced materials is that its components act together as a single anisotropic unit upto the elastic limit, i.e., the two components are bound together so that there can be no relative displacement between them. Many elastic fiber-reinforced composite materials are strongly anisotropic in behavior. Fibre-reinforced materials are a family of composite materials, where the polymer fibers are reinforced by highly oriented polymer fibers, derived from the same fiber. Alumina or concrete is an example of fibre-reinforced material. Under certain temperatures and pressures some fiber materials may also be modified to self-reinforced materials by reinforcing a matrix material of the same fibre. In real life, the fibres might be carbon, nylon, or, conceivably metal whiskers. It is important to study the interactions between individual fibres and the surrounding matrix for understanding the mechanical properties of fibre-reinforced materials inside the earth. Reinforcement is a part of composite material that provides strength, stiffness, and the ability to carry a load. The idea of continuous self-reinforcement at every point of an elastic solid was introduced by Belfield et al. [10]. When these types of rocks come in the way of seismic waves, they influence their propagation. So, the propagation of SH-wave in anisotropic elastic media is of considerable importance. To investigate the effect of reinforcement on the propagation of magnetoelastic shear waves (SH-waves), crack studies have been carried out by Chattopadhyay et al. [11] and Chattopadhyay et al. [12].

The interest in studying the problem of wave propagation in micropolar elastic medium is because of its applications in many physical substances, e.g., material particles having rigid directors, chopped fiber composites, platelet composites, aluminium epoxy, liquid crystal with side chains, a large class of substances like liquid crystals with rigid molecules, rigid suspensions, animal blood with rigid cells, foams, porous materials, bones, magnetic fluids, clouds with dust, concrete with sand and muddy fluids. The classical theory of elasticity explains the behavior of those materials which are considered as continuum in mathematical sense such as steel, aluminium, concrete, coal etc but it is inadequate to explain the behavior of materials like polymers, cellular solids, and crystals exhibiting microstructure. In particular, this theory doesn't explain some discrepancies that occur in the case of elastic vibration of high frequency and small wave length. It was desired to develop size-dependent consistent continuum mechanics that can analyze the behavior of material at microscale and explain the behavior at macroscale. Voigt [13] was the first who tried to remove the shortcomings of the classical theory of elasticity by the assumption that interaction between two particles of a body through an area element lying with the material is not only transmitted by force vector but also by a couple (moment) vector. This gave rise to couple stresses in elasticity. However, the complete theory of asymmetric elasticity was developed by Cosserat and Cosserat [14]. They assumed that the body consist of interconnected particles in the form of small rigid bodies and deformation of the medium is described not only by the displacement vector but also by independent rotation vector. Further, Eringen [15] generalised the classical theory of elasticity by taking into account that the directors are rigid and there are three rotational degrees of freedom in addition to the three classical displacement degree of freedom. Recently, Altenbach and Eremeyev [16] have highlighted the three-dimensional Cosserat-type model and its possibility to describe complex media like micro-inhomogeneous materials, polycrystalline and cellular solids, foams, lattices, masonries, particle assemblies, magnetic rheological fluids, liquid crystals, etc.

The objective of the present work is to study the problem of Love wave propagation in a vertically homogeneous fibre-reinforced layer welded imperfectly with micropolar elastic half-space. The dispersion equation for the propagation of Love type wave has been obtained in closed form. The study also reveals the influence of imperfect factor associated with common interface, reinforcement parameter associated with layer, micropolarity and coupling factor associated with half-space on the phase velocity of Love wave. Also, some results are derived from our analysis as a particular case.

2.2 FORMULATION OF THE PROBLEM

We consider a fibre-reinforced layer of width H' imperfectly welded with a micropolar elastic half-space. The rectangular Cartesian coordinate system is considered such that x-axis is along the direction of propagation of Love type wave and the z-axis is positive vertically downward, as shown in Fig. 2.1.

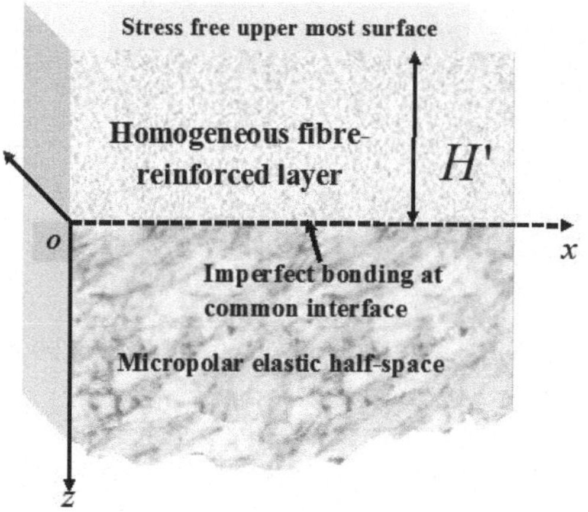

Figure 2.1 Geometry of the problem.

Let us assume (u_1, v_1, w_1) and (u_2, v_2, w_2) as the displacement component caused due to propagation of Love type wave in upper layer and lower half-space respectively. Now, for the propagation of Love type wave in the x-direction and causing displacement only in y-direction, we shall assume that

$$u_i = 0,\ v_i = v_i(x,z,t),\ w_i = 0;\ (i = 1,2). \tag{1}$$

2.2.1 SOLUTION FOR HOMOGENEOUS FIBER-REINFORCED LAYER

The constitutive equations for a fibre-reinforced linearly elastic medium are

$$\tau_{ij} = \lambda' e_{kk}\delta_{ij} + 2\mu_T e_{ij} + \alpha'(a_k a_m e_{km}\delta ij + e_{kk}a_i a_j + 2(\mu_L - \mu_T)(a_i a_k e_{kj} + a_j a_k e_{ki}) \\ + \beta' a_k a_m e_{km} a_i a_j;\ (i,j,k,m = 1,2,3), \tag{2}$$

where τ_{ij} are components of stress, $e_{ij} = \frac{u_{i,j}+u_{j,i}}{2}$ are components of infinitesimal strain, δ_{ij} is the Kronecker delta and a_i are the components of $\vec{a}$, all referred to rectangular cartesian co-ordinates x_i, $\vec{a} = (a_1, a_2, a_3)$ is the preferred directions of reinforcement such that $a_1^2 + a_2^2 + a_3^2 = 1$. The vector may be function of position. Indices take the values 1, 2, 3 and summation convention is employed. The coefficients $\lambda', \mu_T, \alpha', \beta'$, and μ_L are elastic constants with dimension of stress. μ_T can be identified as the shear modulus in transverse shear across the preferred direction, and μ_L as the shear modulus in longitudinal shear in the preferred direction. α' and β' are specific stress components to take into account different layers for concrete part of the composite material. In the absence of body forces, the only non-vanishing equation of motion for small elastic disturbance in fibre-reinforced layer is

$$\frac{\partial \tau_{21}}{\partial x} + \frac{\partial \tau_{23}}{\partial z} = \rho_1 \frac{\partial^2 v_2}{\partial t^2}, \tag{3}$$

where

$$\tau_{21} = \left[\mu_T \frac{\partial v_1}{\partial x} + (\mu_L - \mu_T)a_1\left(a_1 \frac{\partial v_1}{\partial x} + a_3 \frac{\partial v_1}{\partial z}\right)\right] e^{\nu(z+H')}, \tag{4}$$

$$\tau_{23} = \left[\mu_T \frac{\partial v_1}{\partial z} + (\mu_L - \mu_T)a_3\left(a_1 \frac{\partial v_1}{\partial x} + a_3 \frac{\partial v_1}{\partial z}\right)\right] e^{\nu(z+H')}. \tag{5}$$

Using equations (1) together with stresses (4) and (5), the equation of motion (3) for the propagation of Love type wave in fibre-reinforced layer reduces to

$$P'\frac{\partial^2 v_1}{\partial x^2} + Q'\frac{\partial^2 v_1}{\partial x \partial z} + R'\frac{\partial^2 v_1}{\partial z^2} + S'\frac{\partial v_1}{\partial x} + T'\frac{\partial v_1}{\partial z} = \rho_1 \frac{\partial^2 v_1}{\partial t^2}, \tag{6}$$

where
$P' = \mu_T + a_1^2(\mu_L - \mu_T)$, $Q' = 2a_1a_3(\mu_L - \mu_T)$, $R' = \mu_T + a_3^2(\mu_L - \mu_T)$,
$S' = \nu a_1 a_3(\mu_L - \mu_T)$ and $T' = \nu(\mu_L + a_3^2(\mu_L - \mu_T))$.
Solution of (6) may be assumed as

$$v_1 = V_1(z)e^{ik(x-Vt)}, \tag{7}$$

where k is the wave number and V is the phase velocity of Love type wave.
In view of (7), equation (6) gives,

$$R'\frac{d^2V_1}{dz^2} + (ikQ' + T')\frac{dV_1}{dz} + (ikS' - P'k^2 + \rho_1 k^2 V^2)V_1(z) = 0. \tag{8}$$

Solution of equation (8) is

$$v_1(z) = (Ae^{m_1 z} + Be^{m_2 z})e^{ik(x-Vt)}, \tag{9}$$

where m_1 and m_2 are given in Appendix 1.

2.2.2 SOLUTION FOR MICROPOLAR ELASTIC HALF-SPACE

The equation of motion (Eringen [15]) for micropolar elastic half-space in the absence of body force may be written as:

$$(\lambda+\mu)\nabla(\nabla.\vec{u})+(\mu+\kappa)\nabla^2\vec{u}+\kappa(\nabla\times\vec{\phi})=\rho_2\frac{\partial^2\vec{u}}{\partial t^2}, \tag{10}$$

$$(\alpha+\beta+\gamma)\nabla(\nabla.\vec{\phi})-\gamma\nabla\times(\nabla\times\vec{\phi})+\kappa(\nabla\times\vec{u})-2\kappa\vec{\phi}=\rho_2 j\frac{\partial^2\vec{\phi}}{\partial t^2}. \tag{11}$$

With the constitutive relations

$$\sigma_{in}=\lambda u_{l,l}\delta_{in}+\mu(u_{i,n}+u_{n,i})+\kappa(u_{n,i}-\varepsilon_{inl}\phi_l), \tag{12}$$

$$m_{in}=\alpha\phi_{l,l}\delta_{in}+\beta\phi_{i,n}+\gamma\phi_{n,i}, \tag{13}$$

where $i,n,l=1,2,3$, $\vec{u}=(u_2,v_2,w_2)$, $\vec{\phi}=(\phi_1,\phi_2,\phi_3)$, λ, μ are Lame parameter and $\gamma,\kappa,\alpha,\beta$ are micropolar material constants, ρ_2 is the density. σ_{in} and m_{in} are the stress tensor and couple stress tensor, respectively, and δ_{in} is the Kronecker delta, ε_{inl} is permutation tensor and j is the micro-inertia.

Let us introduce ψ and φ defined by

$$\phi_1=\frac{\partial\psi}{\partial x}+\frac{\partial\varphi}{\partial z}\text{ and }\phi_3=\frac{\partial\psi}{\partial z}-\frac{\partial\varphi}{\partial x}. \tag{14}$$

Using equations (1) and (14), the system of equation (10) and (11) reduces for the propagation of Love type wave as

$$\nabla^2 v_2+c_1^2\nabla^2\varphi=\frac{1}{c_2^2}\frac{\partial^2 v_2}{\partial t^2}, \tag{15}$$

$$\nabla^2\psi-2\frac{c_5^2}{c_3^2+c_4^2}\psi=\frac{1}{c_3^2+c_4^2}\frac{\partial^2\psi}{\partial t^2}, \tag{16}$$

$$\nabla^2\varphi-2\frac{c_5^2}{c_3^2}\varphi-\frac{c_5^2}{c_3^2}v_2=\frac{1}{c_3^2}\frac{\partial^2\varphi}{\partial t^2}, \tag{17}$$

where c_1,c_2,c_3,c_4,c_5 are given in Appendix 1.

We assume solution of the form

$$(\psi,\varphi,v_2)(x,z,t)=(\psi,\varphi,v_2)(z)e^{ik(x-Vt)}. \tag{18}$$

With the help of (18), equations (15), (16), and (17) results in

$$\frac{d^2\psi}{dz^2}+r^2\psi=0, \tag{19}$$

$$[D^4+PD^2+Q](\varphi,v_2)=0. \tag{20}$$

where $D\equiv\frac{d}{dz}$, p, q, and r are given in Appendix 1.

The solution of equations (19) and (20) leads to

$$v_2=(Fe^{-ipz}s_1+He^{-iqz}s_2)e^{ik(x-Vt)}, \tag{22}$$

$$\phi_1=(ikD'e^{-irz}-ipFe^{-ipz}-iqHe^{-iqz})e^{ik(x-Vt)}, \tag{23}$$

$$\phi_3=[-irD'e^{-irz}-ik(Fe^{-ipz}+He^{-iqz})]e^{ik(x-Vt)}, \tag{24}$$

where D',F,H are the arbitrary constants, s_1 and s_2 are given in Appendix 1.

2.3 BOUNDARY CONDITIONS

For the propagation of Love type wave in vertically homogeneous fibre-reinforced layer imperfectly welded with a micropolar elastic half-space, the following boundary conditions are to be satisfied

(i) The upper most surface of vertically homogeneous fiber-reinforced layer is stress free, i.e.,

$$\tau_{yz} = 0 \text{ at } z = -H'. \tag{25}$$

(ii) Layer and half-space are imperfectly welded, i.e.,

$$\tau_{yz} = L(v_2 - v_1) \text{ at } z = 0, \tag{26}$$

where L is the degree of imperfectness of common interface.

(iii) Stresses are continuous at the common interface of layer and half-space, i.e.

$$\tau_{yz} = \sigma_{yz} \text{ at } z = 0. \tag{27}$$

(iv) Couple stresses vanishes at the common interface of layer and half-space, i.e.

$$\begin{aligned} m_{zz} &= 0 \text{ at } z = 0, \\ m_{zx} &= 0 \text{ at } z = 0. \end{aligned} \tag{28}$$

Using equations (7), (22), (23), and (24) together with the boundary conditions (25), (26), (27), and (28), we get

$$e^{-m_1 H'} s_3 A + e^{-m_2 H'} s_4 B = 0, \tag{29}$$

$$e^{\nu H'}(s_3 A + s_4 B) = i\kappa k D' - i s_5 F - i s_6 H, \tag{30}$$

$$(e^{\nu H'} s_3 + L)A + (e^{\nu H'} s_4 + L)B = F L s_1 + H L s_2, \tag{31}$$

$$s_5 D' + s_6 F + s_7 H = 0, \tag{32}$$

$$s_8 D' + s_9 F + s_{10} H = 0, \tag{33}$$

where $s_i(i = 3, ..., 10)$ are provided in Appendix 2.
Now, elimination of arbitrary constants A, B, D', F and H from the equations (29), (30), (31), (32) and (33) leads to

$$\tan(m_6 H') = \frac{\xi_3 \cos(m_4 H') + \xi_4 \sin(m_4 H') - \xi_5 \cos(m_6 H')}{\xi_6 \cos(m_6 H')}, \tag{34}$$

where ξ_3, ξ_4, ξ_5, and ξ_6 are given in Appendix 2.
Equation (34) is the dispersion equation for Love type wave in a vertically homogeneous fiber-reinforced layer imperfectly welded with a micropolar elastic half-space.

2.4 PARTICULAR CASES

2.4.1 CASE 1

To describe interface damping, we use a complex interface stiffness with an imaginary part. Given $L = L_1 + iL_2$, where both L_1 and L_2 are real. Hence,

$$\Gamma = \Gamma_1 - i\Gamma_2 = \frac{\mu k}{|L|^2}(L_1 - iL_2),$$

where Γ_1 and Γ_2 are real. Γ_1 corresponds to the flexibility parameter of imperfect interface, whereas Γ_2 is the viscoelastic parameter of imperfect interface.

When $L = L_1 + iL_2$, dispersion equation (34) reduces to

$$\tan(m_6 H') = \frac{\chi_{13}\chi_{15} + \chi_{14}\chi_{16}}{\chi_{15}^2 + \chi_{16}^2}, \tag{35}$$

where $\chi_{13}, \chi_{14}, \chi_{15}$ and χ_{16} are given in Appendix 3.
Equation (35) is dispersion equation for the propagation of Love type wave in vertically homogeneous fibre-reinforced layer lying over a micropolar elastic half-space when interface is complex.

2.4.2 CASE 2

When $L \to \infty$ or $\Gamma \to 0$ dispersion equation (34) reduces to

$$\tan(m_6 H') = \frac{\chi_3' \cos(m_4 H') + \chi_4' \sin(m_4 H') - \chi_5' \cos(m_6 H')}{\chi_6' \cos(m_6 H')}, \tag{36}$$

where $\chi_3', \chi_4', \chi_5'$, and χ_6' are given in Appendix 3.
Equation (36) is dispersion equation for the propagation of Love type wave in vertically homogeneous fibre-reinforced layer perfectly welded with micropolar elastic half-space.

2.4.3 CASE 3

When $\mu_L = \mu_T = \mu_1 0$ and $a_1 = a_3 = 0$ dispersion equation (34) reduces to

$$\tan(m_6'' H') = \frac{\xi_3'' \cos(m_4'' H') + \xi_4'' \sin(m_4'' H') - \xi_5'' \cos(m_6'' H')}{\xi_6'' \cos(m_6'' H')}, \tag{37}$$

where m_4'', m_6'', ξ_3'', ξ_4'', ξ_5'', and ξ_6'' are given in Appendix 4.
Equation (37) is dispersion equation for the propagation of Love type wave in vertically homogeneous reinforced-free layer imperfectly welded with micropolar elastic half-space.

2.5 CONCLUSION

In the present paper propagation of Love type wave in a fibre-reinforced lying over a micropolar elastic half-space has been studied. Dispersion equation has been established in closed form and the effects of micropolarity associated with half-space and imperfectness of common interface on the phase velocity of Love type wave have been observed. Following points can be outlined as an outcome of the study:

- Wave number affects the phase velocity of Love type wave substantially. In reinforced layer lying over a micropolar elastic half-space phase velocity decreases with increases in wave number.
- Substantial effect of micropolar parameter on phase velocity of Love type wave. With increasing value of micropolar parameter phase velocity decreases.
- It is observed that micropolarity has remarkable effect on phase velocity of Love type wave. Phase velocity of Love type wave increases with increase in coupling factor in both the considered cases.
- Imperfectness associated with common interface has significant effect on phase velocity of Love type wave. Imperfectness factor is taken as real and complex. Real imperfectness factor and viscoelastic parameter of common imperfect interface disfavors the phase velocity of Love type wave whereas flexibility parameter of imperfect common interface favors the phase velocity of Love type wave.
- Reinforcement parameter of layer disfavors phase velocity of Love type wave, i.e., as reinforcement prevails in the medium phase velocity decreases.

APPENDIX 1

$m_1 = m_3 + im_4,\ m_2 = m_5 + im_6,\ m_3 = \dfrac{-T' - r'^{1/2}sin(\frac{\theta}{2})}{2R'},\ m_4 = \dfrac{r'^{1/2}cos(\frac{\theta}{2}) - kQ'}{2R'},\ m_5 =$
$\dfrac{-T' + r'^{1/2}sin(\frac{\theta}{2})}{2R'}, m_6 = \dfrac{-r'^{1/2}cos(\frac{\theta}{2}) - kQ'}{2R'},$

$$\theta = tan^{-1}\left(\frac{4R'S'k - 2kQ'T'}{k^2Q'^2 + 4R'\rho_1^0k^2V^2 - T'^2 - 4R'P'k^2}\right),$$

$$r' = \sqrt{(4R'S'k - 2kQ'T')^2 + (k^2Q'^2 + 4R'\rho_1^0k^2V^2 - T'^2 - 4R'P'k^2)^2},$$

$$c_1 = \sqrt{\frac{\kappa}{\mu+\kappa}}, c_2 = \sqrt{\frac{\mu+\kappa}{\rho_2}}, c_3 = \sqrt{\frac{\gamma}{\rho_2 j}}, c_4 = \sqrt{\frac{\alpha+\beta}{\rho_2 j}}, c_5 = \sqrt{\frac{\kappa}{\rho_2 j}},$$

$$p^2, q^2 = \frac{-P \pm \sqrt{P^2 - 4Q}}{2}, r^2 = \left(\frac{k^2V^2}{c_3^2 + c_4^2} - \frac{2c_5^2}{c_3^2 + c_4^2} - k^2\right),$$

$$P = -\frac{c_5^2}{c_3^2}(c_1^2 - 2) + k^2V^2\left(\frac{1}{c_3^2} + \frac{1}{c_2^2}\right) - 2k^2,$$

$$Q = \left(\frac{k^2V^2}{c_2^2} - k^2\right)\left(\frac{k^2V^2}{c_3^2} - \frac{2c_5^2}{c_3^2} - k^2\right) - k^2\frac{c_1c_5^2}{c_3^2},$$

$$s_1 = \left(-p^2 + \frac{k^2V^2}{c_3^2} - \frac{2c_5^2}{c_3^2} - k^2\right)\frac{c_3^2}{c_5^2}, s_2 = \left(-q^2 + \frac{k^2V^2}{c_3^2} - \frac{2c_5^2}{c_3^2} - k^2\right)\frac{c_3^2}{c_5^2}.$$

APPENDIX 2

$s_3 = t_1 + it_2$, $t_1 = m_3\mu_L + m_3a_3^2(\mu_L - \mu_T)$, $t_2 = m_4\mu_T + a_1a_3(\mu_L - \mu_T)k + m_4a_3^2(\mu_L - \mu_T)$, $s_4 = t_3 + it_4$, $t_3 = m_5\mu_t + m_5a_3^2(\mu_L - \mu_T)$, $t_4 = m_6\mu_T + a_1a_3(\mu_L - \mu_T)k + m_6a_3^2(\mu_L - \mu_T)$, $s_5 = \mu s_1 + \kappa$, $s_6 = \mu s_2 + \kappa$, $s_7 = -\alpha(k^2 - r^2) - r^2(\beta+\gamma)$, $s_8 = -kp(\beta+\gamma)$, $s_9 = -kq(\beta+\gamma)$, $s_{10} = (-\beta k^2 + \gamma rk)$,
$s_{11} = (-kr\beta - p^2\gamma)$, $s_{12} = (-kq\beta - q^2\gamma)$,
$s_{13} = s_1(s_8s_{12} - s_{11}s_9) - s_2(s_7s_{11} - s_{10}s_8)$,
$s_{14} = \kappa k(s_8s_{12} - s_9s_{11}) - s_5(s_7s_{12} - s_9s_{10}) + s_6(s_7s_{11} - s_8s_{10})$,
$\xi_1 = (t_1t_4 + t_2t_3)(Ls_{13} + s_{14})e^{\nu H'}$, $\xi_2 = (t_1t_3 - t_2t_4)(Ls_{13} + s_{14})e^{\nu H'}$,
$\xi_3 = (\xi_2 + Lt_1s_{14})e^{-m_3H'}$, $\xi_4 = (\xi_1 + Lt_2s_{14})e^{-m_3H'}$,
$\xi_5 = (\xi_2 + Lt_3s_{14})e^{-m_5H'}$, $\xi_6 = (\xi_1 + Lt_4s_{14})e^{-m_5H'}$.

APPENDIX 3

$\chi_1 = (t_1t_4 + t_2t_3)(L_1s_{13} + s_{14})e^{\nu H'}$, $\chi_2 = (t_1t_4 + t_2t_3)L_2s_{13}e^{\nu H'}$,
$\chi_3 = (t_1t_3 - t_2t_4)(L_1s_{13} + s_{14})e^{\nu H'}$, $\chi_4 = (t_1t_3 - t_2t_4)L_1s_{13}e^{\nu H'}$,
$\chi_5 = (\chi_3 + L_1t_1s_{14})e^{-m_3H'}$, $\chi_6 = (\chi_4 + L_2t_1s_{14})e^{-m_3H'}$, $\chi_7 = (\chi_1 + L_1t_2s_{14})e^{-m_3H'}$,
$\chi_8 = (\chi_2 + L_2t_2s_{14})e^{-m_3H'}$, $\chi_9 = (\chi_3 + L_1t_3s_{14})e^{-m_5H'}$, $\chi_{10} = (\chi_4 + L_2t_3s_{14})e^{-m_5H'}$,
$\chi_{11} = (\chi_1 + L_1t_4s_{14})e^{-m_5H'}$, $\chi_{12} = (\chi_2 + L_2t_4s_{14})e^{-m_5H'}$,
$\chi_{13} = \chi_5\cos(m_4H') + \chi_7\sin(m_4H') - \chi_9\cos(m_6H')$,
$\chi_{14} = \chi_6\cos(m_4H') + \chi_8\sin(m_4H') - \chi_{10}\cos(m_6H')$,
$\chi_{15} = \chi_{11}\cos(m_6H')$, $\chi_{16} = \chi_{12}\cos(m_6H')$, $\chi_3' = t_1s_{14}e^{-m_3H'}$,
$\chi_4' = t_2s_{14}e^{-m_3H'}$, $\chi_5' = t_3s_{14}e^{-m_5H'}$, $\chi_6' = t_4s_{14}e^{-m_5H'}$.

APPENDIX 4

$$r_1' = 4\mu_1\rho_1k^2V^2 - v^2\mu_1{}^2 - 4\mu_1{}^2k^2,\ m_3'' = -\frac{v}{2},\ m_4'' = \frac{r_1'^{1/2}}{2\mu_1},$$

$$m_5'' = -\frac{v}{2},\ m_6'' = -\frac{r_1'^{1/2}}{2\mu_1},\ t_1'' = -\frac{v\mu_1}{2},\ t_2'' = m_4''\mu_1,\ t_3'' = -\frac{v\mu_1}{2},\ t_4'' = m_6''\mu_1,$$

$$\xi_1'' = (t_1''t_4'' + t_2''t_3'')(Ls_{13} + s_{14})e^{vH'},\ \xi_2'' = (t_1''t_3'' - t_2''t_4'')(Ls_{13} + s_{14})e^{vH'},$$

$$\xi_3'' = (\xi_2'' + Lt_1''s_{14})e^{-m_3H'},\ \xi_4'' = (\xi_1'' + Lt_2''s_{14})e^{-m_3H'},$$

$$\xi_5'' = (\xi_2'' + Lt_3''s_{14})e^{-m_5H'},\ \xi_6'' = (\xi_1'' + Lt_4''s_{14})e^{-m_5H'}.$$

REFERENCES

1. Jones JP and Whittier JS (1967) Waves in a flexible bounded interface. Journal of Applied Mechanics 34: 905-909.
2. Murty GS (1975) A theoretical model for the attenuation and dispersion of stonely waves at the loosely bounded interface of elastic half-spaces. Physics of the Earth and Planetary Interiors 11: 65-79.
3. Nayfeh AH and Nassar EM (1978) Simulation of the influence of bonding materials on the dynamic behaviour of laminated composites. Journal of Applied Mechanics 45: 822-828.
4. Lavrentyev AI and Rokhlin SI (1998) Ultrasonic spectroscopy of imperfect contact interfaces between a layer and two solids. Journal of the Acoustical Society of America 103(2): 657-664.
5. Fan H and Sze KY (2001) A micro-mechanics model for imperfect interface in dielectric materials. Mechanics of Materials. 33: 363-370.
6. Wang X and Zhong Z (2003) Three-dimensional solution of smart laminated anisotropic circular cylindrical shellswith imperfect bonding. International Journal of Solid and Structures 40: 5901-5921.
7. Shodja HM, Tabatabaei SM and Kamali HT (2006) A piezoelectric-inhomogenity system with imperfect interface. International Journal of Engineering Science 44: 291-311.
8. Shariat BAS and Eslami MR (2006) Thermal buckling of imperfect functionally graded plates. International Journal of Solid and Structures 43: 4082-4096.
9. Peng Li and Feng Jin (2015) Excitation and propagation of shear horizontal waves in a piezoelectric layer imperfectly bonded to a metal or elastic substrate. Acta Mechinca 226(2): 267-284.
10. Belfield AJ, Rogers TG and Spencer AJM (1983) Stress in elastic plates reinforced by fibres lying in concentric circles, Journal of the Mechanics and Physics of Solids 31(1): 25-54.
11. Chattopadhyay A, Gupta S, Chattopadhyay A and Singh AK (2010) The dispersion of shear wave in multilayered magnetoelastic selfreinforced media. International Journal of Solids and Structures 47(9): 1317-1324.
12. Chattopadhyay A, Singh AK and Dhua S (2014) Effect of heterogeneity and reinforcement on propagation of a crack due to shear waves. International Journal of Geomechanics 14(4): 04014013.
13. Voigt W (1887) Theoretishe studien uber die ebsti-Zitatsverhaltnisse der Kristalle. Abh Ges Wiss Gottingen 34.
14. Cosserat E and Cosserat F (1909) Theorie des corps deformable. Hermann. Paris.
15. Eringen AC (1966) Linear theory of micropolar elasticity. Journal of Mathematics and Mechanics 15: 909-923.
16. Altenbach H and Eremeyev VA (2013) Generalized Continua - from the Theory to Engineering Applications. DOI:10.1007/978-3-7091-1371-4.
17. Chattopadhyay A and Chaudhury S (1995) Magnetoelastic shear waves in an infinite self- reinforced plate. International Journal for Numerical and Analytical Methods in Geomechanics 19(4): 289–304.
18. Gauthier RD (1982) Experimental investigation on micropolar media, In: Brulin O, Hsieh RKT (eds) Mechanics of micropolar media. World Scientific. Singapore: 395-463.
19. Gubbins D (1990) Seismology and Plate Tectonics. Cambridge. Cambridge University Press: 170.
20. Fatemi J, Van KF and Onck PR (2002) Generalized continuum theories: application to stress analysis in bone. Meccanica 37: 385-396.

3 On Induced Map $\eta^{\#}$ in Fuzzy Set Theory and Its Applications

Sandeep Kaur
Department of Mathematics and Statistics,
Central University of Punjab, Bathinda, Punjab, India

Nitakshi Goyal
Department of Mathematics,
Akal Degree College, Mastuana Sahib (Sangrur), India

CONTENTS

In this study, we introduce the # image of a fuzzy set which gives an induced map $\eta^{\#}$ corresponding to any function $\eta : U \to V$, where U and V are crisp sets. With this, we present a new vision of studying fuzzy continuous mappings in fuzzy topological spaces where fuzzy continuity explains the term of closeness in the mathematical models. We also define the concept of fuzzy saturated sets which helps us to prove some new characterizations of fuzzy continuous mappings in terms of the interior operator rather than the closure operator.

3.1 INTRODUCTION

The theory of fuzzy sets, which offers to deal with uncertainty, vagueness, and imprecise boundaries was introduced by L.A. Zadeh in [8]. Since its advent, research in fuzzy set theory and its applications in several directions has been expanding exponentially which results in significant success from conceptual and technological viewpoints. Therefore, fuzzification of various classical concepts related to topology has been done by various authors in [1], [7], [4], etc. Chang introduced the picture of fuzzy topological spaces in [3] and generalized the notion of continuity to what he named fuzzy continuous (F-continuous) functions and studied some of its characterizations. To introduce fuzzy continuity, he introduced an image and inverse image of a fuzzy set corresponding to any function $\eta : U \to V$ and established a few properties of fuzzy sets induced by mappings.

In [4], Ming et al. investigated the theory of interior and closure of a fuzzy set in a fuzzy topological space and studied some of its properties which are similar to that in general topological

DOI: 10.1201/9781032634142-3

space. They also studied some more characterizations of fuzzy continuous functions in their other papers [5]. In [2], Arkhangel'skii introduced a new map $\eta^{\#} : \wp(U) \to \wp(V)$ given by $\eta^{\#}(M) = \{v \in V | \eta^{-1}(v) \subset M\}$ for any subset M of U, where $\wp(U)$ is a collection of all subsets of U. This gives us the idea about the possibility of introducing such a type of induced map using fuzzy subsets and using it to find possible characterizations of fuzzy continuity.

In this paper, apart from the image and inverse image of a fuzzy set, we introduce # image of a fuzzy set. Equivalently, given any function $\eta : U \to V$, where U and V are crisp sets, it gives rise to an induced map $\eta^{\#} : FZ(U) \to FZ(V)$, whereby $FZ(U)$, we mean a collection of all fuzzy subsets of U (Definition 3.5). We prove the monotonicity of map $\eta^{\#}$ along with its various properties (Lemma 3.1). The relationship between the # image of a fuzzy set and the image of a fuzzy set is also given (Lemma 3.2). Further, we define a fuzzy saturated set and obtain its characterizations. It is also shown that surjective maps can be characterized in terms of # image of a fuzzy set (Theorem 3.3). Finally, using $\eta^{\#}$ map, we give characterizations of fuzzy continuous maps in terms of the interior operator rather than the closure operator (Theorem 3.4) and using saturated sets (Theorem 3.5).

3.2 PRELIMINARIES

In this segment, we will present a few basic definitions and consequences that we need in our further sections.

Let $U = \{a\}$ be a space of points then a *fuzzy set M* in U is a function μ_M from U to closed interval $[0,1]$, i.e., $\mu_M : U \to [0,1]$. To rephrase it, a fuzzy set is specified by a membership mapping that connects with each u in U its "grade of membership" $\mu_M(u)$, in M. An empty fuzzy set, denoted by ϕ is described as $\mu_\phi(a) = 0$ for all a in U and fuzzy set U is defined as $\mu_U(a) = 1$ for all a in U. For a fuzzy set M,

$$Hgt(M) = \sup\{\mu_M(a) | a \in U\}$$

is known as *height* of M and a fuzzy set M is said to be *normalized* if and only if $Hgt(M) = 1$. Also, two fuzzy sets are equal if and only if their membership values at every point of U are equal. For fuzzy sets M, N, G in a space U [3], $G = M \cup N$ if and only if $\mu_G(a) = max\{\mu_M(a), \mu_N(a)\}$, $G = M \cap N$ if and only if $\mu_G(a) = min\{\mu_M(a), \mu_N(a)\}$ and $G = M^c$ if and only if $\mu_G(a) = 1 - \mu_M(a)$, for all a in U.

Definition 3.1 *[3] "A family T of fuzzy sets in U is said to be a fuzzy topology if it fulfills the following conditions:*

(a) ϕ and $U \in T$.
(b) $N \cap G \in T$ for $N, G \in T$.
(c) $\cup N_i \in T$ for $N_i \in T$, $i \in I$.

Members of T are known as T-open fuzzy sets (or fuzzy open sets) and the pair (U,T) is known as a fuzzy topological space (or fts). Also, a fuzzy set is T-closed (or fuzzy closed) if and only if its complement is T-open."

Definition 3.2 *[4] "For a fuzzy topological space (U,T) and a fuzzy set G in U, interior of G is the union of all T-open fuzzy sets contained in G, denoted by G^o and closure of G is the smallest T-closed fuzzy set containing G, denoted by $\overline{G}$."*

Theorem 3.1 *[4] "In a fts (U,T), for any fuzzy subset M of U, $(\overline{M})^c = (M^c)^o$ and so $\overline{M^c} = (M^o)^c$."*

Definition 3.3 *[3] "Let $h : U \to V$ be a mapping and G be a fuzzy set in U then the image of G, denoted by $h(G)$ is a fuzzy set in V where its membership mapping is defined as,*

$$\mu_{h(G)}(v) = \sup_{z \in h^{-1}(v)} \mu_G(z) \qquad if\ h^{-1}(v) \neq \emptyset$$
$$= \quad 0, \qquad otherwise.$$

for all $v \in V$ where $h^{-1}(v) = \{u \mid h(u) = v\}$.

Also for a fuzzy set K in V with membership function $\mu_K(v)$, inverse image of K, denoted by $h^{-1}(K)$ is a fuzzy set in U where its membership mapping is defined as

$$\mu_{h^{-1}(K)}(u) = \mu_K(h(u))$$

for all u in U."

A few properties of the above fuzzy mapping are given in [3]

Definition 3.4 *[4] "A function $\eta : (U,S) \to (V,T)$ is fuzzy continuous (or F-continuous) iff the inverse image of each T-open fuzzy set is S-open fuzzy set."*

Theorem 3.2 *[5] "A function $\eta : (U,S) \to (V,T)$ is fuzzy continuous (or F-continuous) iff $\eta(\overline{M}) \subset \overline{\eta(M)}$, for any fuzzy set M in U."*

3.3 INDUCED MAP $\eta^{\#}$

We begin by introducing # image of a fuzzy set which gives rise to the induced map $\eta^{\#} : FZ(U) \to FZ(V)$ corresponding to any function $\eta : U \to V$.

Definition 3.5 *Let $U = \{u\}$ and $V = \{v\}$ be spaces of points. Let η be a function from U to V and M be a fuzzy set in U with membership map $\mu_M(u)$. Then # image of a fuzzy set M, written as $\eta^{\#}(M)$, is a fuzzy set in V with a membership function defined by*

$$\mu_{\eta^{\#}(M)}(v) = \inf_{z \in \eta^{-1}(v)} \mu_M(z) \qquad if\ \eta^{-1}(v) \neq \emptyset$$
$$= \quad 1, \qquad otherwise.$$

for all $v \in V$ where $\eta^{-1}(v) = \{u \mid \eta(u) = v\}$.

Remark 3.1 *(1) If $\eta : U \to V$ is surjective map then $\eta^{\#}(M) \subset \eta(M)$ for every fuzzy subset M in U, since $\eta^{-1}(v) \neq \emptyset$ for all $v \in V$.*
(2) If $\eta : U \to V$ is injective map then $\eta(M) \subset \eta^{\#}(M)$ for every fuzzy subset M in U, since

$$\mu_{\eta^{\#}(M)}(v) = \inf_{u \in \eta^{-1}(v)} \mu_M(u) = \mu_M(u) =$$
$$\sup_{u \in \eta^{-1}(v)} \mu_M(u),$$

where $\eta^{-1}(v) = u$ for some $v \in V$.

Note: From the definition of $\eta^{\#}$ map, it can be seen easily that # image of any fuzzy set is normalized fuzzy set for any map $\eta : U \to V$, which is not surjective. But if η is surjective then the # image of a fuzzy set need not be normalized fuzzy set.

Definition 3.6 *Let $\eta : U \to V$ be a function. Define a fuzzy set $M^{\#} = \eta^{-1}(\eta^{\#}(M))$ in U with membership function as*

$$\mu_{M^{\#}}(u) = \mu_{\eta^{-1}(\eta^{\#}(M))}(u) = \mu_{\eta^{\#}(M)}(\eta(u))$$

The following Lemma gives some properties of $\eta^{\#}$ map.

Lemma 3.1 *Let $\eta : U \to V$ be any function and M, N be fuzzy subsets of U and B be a fuzzy subset of V. Then*

(a) If $M \subset N$ then $(\eta^{\#}(M)) \subset (\eta^{\#}(N))$.
(b) $\eta^{-1}(\eta^{\#}(M)) \subset M$ i.e. $M^{\#} \subset M$ and equality holds if grade of membership $\mu_M(z)$ is constant on fibers.
(c) $B \subset \eta^{\#}(\eta^{-1}(B))$ and equality holds if η is surjective.
(d) $\eta(M^{\#}) = \eta^{\#}(M) \cap \eta(U)$.
(e) $\eta^{\#}(M \cap N) = \eta^{\#}(M) \cap \eta^{\#}(N)$
(f) $\eta^{\#}(\phi) = (\eta(U))^{c}$.
(g) $\eta^{\#}(U) = V$.
(h) $U^{\#} = U$ and $\phi^{\#} = \phi$.
(i) $\eta(M^{\#}) = \eta^{\#}(M) \cap \eta(M)$.
(j) $\eta^{\#}(\eta^{-1}(\eta^{\#}(M))) = \eta^{\#}(M^{\#}) = \eta^{\#}(M)$
(k) $\eta^{-1}(\eta^{\#}(\eta^{-1}(B))) = \eta^{-1}(B)$.

Proof

(a) Let $M \subset N$ then $\mu_M(z) \leq \mu_N(z)$ for all $u \in U$. Therefore, for any $v \in V$ such that $\eta^{-1}(v) \neq \emptyset$

$$\begin{aligned}\mu_{\eta^{\#}(M)}(v) &= \inf_{z \in \eta^{-1}(v)} \mu_M(z)\\ &\leq \inf_{z \in \eta^{-1}(v)} \mu_N(z)\\ &= \mu_{\eta^{\#}(B)}(v)\end{aligned}$$

and if $\eta^{-1}(v) = \emptyset$ then obviously $\mu_{\eta^{\#}(M)}(v) = \mu_{\eta^{\#}(N)}(v)$ and so $\eta^{\#}(M) \subset \eta^{\#}(N)$.

(b) For all $u \in U$,

$$\begin{aligned}\mu_{\eta^{-1}(\eta^{\#}(M))}(u) &= \mu_{\eta^{\#}(M)}(\eta(u))\\ &= \inf_{z \in \eta^{-1}(\eta(u))} \mu_M(z)\\ &\leq \mu_M(u)\end{aligned}$$

Therefore, $\eta^{-1}(\eta^{\#}(M)) = M^{\#} \subset M$.

Further, if grade of membership $\mu_M(z)$ is constant on fibers, i.e., for all z in $\eta^{-1}(\eta(u))$, $\mu_M(z)$ remains same then

$$\inf_{z \in \eta^{-1}(\eta(u))} \mu_M(z) = \mu_M(u)$$

Therefore, $\mu_{\eta^{-1}(\eta^{\#}(M))}(u) = \mu_M(u)$ for each u in U and so $\eta^{-1}(\eta^{\#}(M)) = M$. Hence, in particular, if η is injective then $\eta^{-1}(\eta^{\#}(M)) = M$.

(c) *For each* $v \in V$, *such that* $\eta^{-1}(v) \neq \emptyset$, *then*

$$\begin{aligned}\mu_{\eta^{\#}(\eta^{-1}(B))}(v) &= \inf_{z\in\eta^{-1}(v)} \mu_{\eta^{-1}(B)}(z) \\ &= \inf_{z\in\eta^{-1}(v)} \mu_B(\eta(z)) \\ &= \inf_{z\in\eta^{-1}(v)} \mu_B(v) \\ &= \mu_B(v)\end{aligned}$$

and if $\eta^{-1}(v) = \emptyset$ *then*

$$\mu_B(v) \leq 1 = \mu_{\eta^{\#}(\eta^{-1}(B))}(v)$$

Therefore, $B \subset \eta^{\#}(\eta^{-1}(B))$. *In particular, it can be seen easily that if* η ***is surjective map then*** $B = \eta^{\#}(\eta^{-1}(B))$.

(d) *For each* $v \in V$, *if* $\eta^{-1}(v) \neq \emptyset$, *then*

$$\begin{aligned}\mu_{\eta(U)}(v) &= \sup_{z\in\eta^{-1}(v)} \mu_U(z) \\ &= \sup_{z\in\eta^{-1}(v)} 1 \\ &= 1\end{aligned}$$

and if $\eta^{-1}(v) = \emptyset$, *then*

$$\begin{aligned}\mu_{\eta^{\#}(M)}(v) &= 1 \\ \mu_{\eta(U)}(v) &= 0\end{aligned}$$

Therefore, we have

$$\begin{aligned}\mu_{\eta^{\#}(M)\cap\eta(U)}(v) &= min.\{\mu_{\eta^{\#}(M)}(v), \mu_{\eta(U)}(v)\} \\ &= \begin{cases} \mu_{\eta^{\#}(M)}(v) & \eta^{-1}(v) \neq \emptyset \\ 0 & \eta^{-1}(v) = \emptyset \end{cases}\end{aligned}$$

and

$$\begin{aligned}\mu_{\eta(M^{\#})}(v) &= \begin{cases} \sup_{z\in\eta^{-1}(v)} \mu_{M^{\#}}(z) & \eta^{-1}(v) \neq \emptyset \\ 0 & \eta^{-1}(v) = \emptyset \end{cases} \\ &= \begin{cases} \sup_{z\in\eta^{-1}(v)} \mu_{\eta^{\#}(M)}(v) & \eta^{-1}(v) \neq \emptyset \\ 0 & \eta^{-1}(v) = \emptyset \end{cases} \\ &= \begin{cases} \mu_{\eta^{\#}(M)}(v) & \eta^{-1}(v) \neq \emptyset \\ 0 & \eta^{-1}(v) = \emptyset \end{cases}\end{aligned}$$

Hence $\eta(M^{\#}) = \eta^{\#}(M) \cap \eta(U)$.

(e) *For each* $v \in V$, *if* $\eta^{-1}(v) \neq \emptyset$, *then*

$$\begin{aligned}
\mu_{\eta^{\#}(M\cap N)}(v) &= \inf_{z\in\eta^{-1}(v)} \mu_{M\cap N}(z) \\
&= \inf_{z\in\eta^{-1}(v)} \inf\{\mu_M(z), \mu_N(z)\} \\
&= \inf\{\inf_{z\in\eta^{-1}(v)} \mu_M(z), \inf_{z\in\eta^{-1}(v)} \mu_N(z)\} \\
&= \inf\{\mu_{\eta^{\#}(M)}(v), \mu_{\eta^{\#}(N)}(v)\} \\
&= \mu_{\eta^{\#}(M)\cap\eta^{\#}(N)}(v)
\end{aligned}$$

and if $\eta^{-1}(v) = \emptyset$ *then obviously* $\mu_{\eta^{\#}(M\cap N)}(v) = \mu_{\eta^{\#}(M)\cap\eta^{\#}(N)}(v)$. *Hence* $\eta^{\#}(M\cap N) = \eta^{\#}(M)\cap\eta^{\#}(N)$.

(f) *For each* $v \in V$, *if* $\eta^{-1}(v)$ *is not empty, then*

$$\begin{aligned}
\mu_{\eta^{\#}(\phi)}(v) &= \inf_{z\in\eta^{-1}(v)} \mu_{\phi}(z) \\
&= \inf_{z\in\eta^{-1}(v)} 0 \\
&= 0
\end{aligned}$$

and

$$\begin{aligned}
\mu_{\eta(U)^c}(v) &= 1 - \mu_{\eta(U)}(v) \\
&= 1 - \sup_{z\in\eta^{-1}(v)} \mu_U(z) \\
&= 1 - 1 \\
&= 0
\end{aligned}$$

If $\eta^{-1}(v) = \emptyset$ *then* $\mu_{\eta^{\#}(\phi)}(v) = 1$, *and*

$$\begin{aligned}
\mu_{\eta(U)^c}(v) &= 1 - \mu_{\eta(U)}(v) \\
&= 1 - 0 \\
&= 1
\end{aligned}$$

Hence $\eta^{\#}(\phi) = \eta(U)^c$.

(g) *For each* $v \in V$, *if* $\eta^{-1}(v)$ *is not empty, then*

$$\begin{aligned}
\mu_{\eta^{\#}(U)}(v) &= \inf_{z\in\eta^{-1}(v)} \mu_U(z) \\
&= 1 \\
&= \mu_V(v)
\end{aligned}$$

Also if $\eta^{-1}(v) = \emptyset$, *then*

$$\mu_{\eta^{\#}(U)}(v) = 1 = \mu_V(v)$$

(h) For each $u \in U$,

$$\begin{aligned}\mu_{U^{\#}}(u) &= \mu_{\eta^{-1}(\eta^{\#}(U))}(u)\\ &= \mu_{\eta^{\#}(U)}(\eta(u))\\ &= \inf_{z\in\eta^{-1}(\eta(u))} \mu_U(z)\\ &= \inf_{z\in\eta^{-1}(\eta(u))} 1\\ &= 1\\ &= \mu_U(u)\end{aligned}$$

Similarly, $\phi^{\#} = \phi$.

(i) For each $v \in V$, if $\eta^{-1}(v) \neq \emptyset$, then

$$\begin{aligned}\mu_{\eta^{\#}(M)\cap\eta(M)}(v) &= \inf\{\mu_{\eta^{\#}(M)}(v), \mu_{\eta(M)}(v)\}\\ &= \inf\{\inf_{z\in\eta^{-1}(v)} \mu_M(z), \sup_{z\in\eta^{-1}(v)} \mu_M(z)\}\\ &= \inf_{z\in\eta^{-1}(v)} \mu_M(z)\end{aligned}$$

and if $\eta^{-1}(v) = \emptyset$, then

$$\begin{aligned}\mu_{\eta^{\#}(M)}(v) &= 1\\ \mu_{\eta(M)}(v) &= 0\end{aligned}$$

Therefore, we have

$$\begin{aligned}\mu_{\eta^{\#}(M)\cap\eta(M)}(v) &= \begin{cases} \mu_{\eta^{\#}(M)}(v) & \eta^{-1}(v) \neq \emptyset\\ 0 & \eta^{-1}(v) = \emptyset\end{cases}\\ &= \mu_{\eta(M^{\#})}(v)\end{aligned}$$

Hence $\eta(M^{\#}) = \eta^{\#}(M) \cap \eta(M)$.

(j) For each $v \in V$, if $\eta^{-1}(v) \neq \emptyset$, then

$$\begin{aligned}\mu_{\eta^{\#}(\eta^{-1}(\eta^{\#}(M)))}(v) &= \mu_{\eta^{\#}(M^{\#})}(v)\\ &= \inf_{z\in\eta^{-1}(v)} \mu_{M^{\#}}(z)\\ &= \inf_{z\in\eta^{-1}(v)} \mu_{\eta^{\#}(M)}(\eta(z))\\ &= \mu_{\eta^{\#}(M)}(v)\end{aligned}$$

and if $\eta^{-1}(v) = \emptyset$ then

$$\mu_{\eta^{\#}(\eta^{-1}(\eta^{\#}(M)))}(v) = \mu_{\eta^{\#}(M^{\#})}(v) = 1 = \mu_{\eta^{\#}(M)}(v)$$

Hence $\eta^{\#}(\eta^{-1}(\eta^{\#}(M))) = \eta^{\#}(M^{\#}) = \eta^{\#}(M)$.

(k) For each $u \in U$,

$$\begin{aligned}\mu_{\eta^{-1}(\eta^{\#}(\eta^{-1}(B)))}(u) &= \mu_{\eta^{\#}(\eta^{-1}(B))}\eta(u)\\ &= \inf_{z\in\eta^{-1}(\eta(u))} \mu_{\eta^{-1}(B)}(z)\\ &= \inf_{z\in\eta^{-1}(\eta(u))} \mu_{B}\eta(z)\\ &= \mu_B\eta(u)\\ &= \mu_{\eta^{-1}(B)}(u)\end{aligned}$$

Hence $\eta^{-1}(\eta^{\#}(\eta^{-1}(B))) = \eta^{-1}(B)$.

As we have seen in Lemma 3.1(e) that intersection is preserved by $\eta^{\#}$ map. But the following Example shows that union need not be preserved by $\eta^{\#}$ map.

Example 3.1 *Let $U = \{u_1, u_2, u_3, u_4\}$ and $V = \{v_1, v_2, v_3, v_4\}$ be spaces of points. Consider a function $\eta : U \to V$ defined by $\eta(u_1) = v_1$, $\eta(u_2) = v_2$, $\eta(u_3) = v_2$ and $\eta(u_4) = v_3$ and the fuzzy subsets $M = \{(u_1, 0.1), (u_2, 0.4), (u_3, 0.5), (u_4, 0.3)\}$ and $N = \{(u_1, 0.5), (u_2, 0.6), (u_3, 0.3), (u_4, 0.8)\}$ of U. Then $M \cup N = \{(u_1, 0.5), (u_2, 0.6), (u_3, 0.5), (u_4, 0.8)\}$. Now,*

$$\begin{aligned}\mu_{\eta^{\#}(M\cup N)}(v_2) &= \inf_{u\in\eta^{-1}(v_2)} \mu_{M\cup N}(u)\\ &= \inf\{\mu_{M\cup N}(u_2), \mu_{M\cup N}(u_3)\}\\ &= \inf\{0.6, 0.5\}\\ &= 0.5\end{aligned}$$

and

$$\begin{aligned}\mu_{\eta^{\#}(M)\cup\eta^{\#}(N)}(v_2) &= \sup\{\mu_{\eta^{\#}(M)}(v_2), \mu_{\eta^{\#}(N)}(v_2)\}\\ &= \sup\{0.4, 0.3\}\\ &= 0.4\end{aligned}$$

Hence $\eta^{\#}(M \cup N) \neq \eta^{\#}(M) \cup \eta^{\#}(N)$.

The following Lemma gives some more properties of $\eta^{\#}$ map which we shall need to characterize fuzzy continuity.

Lemma 3.2 *Let $\eta : U \to V$ be any function and M be any fuzzy subset of U and B be any fuzzy subset of V. Then*

(a) $\eta^{-1}(B) \subset M$ if and only if $B \subset \eta^{\#}(M)$.
(b) $(g \circ \eta)^{\#} = g^{\#} \circ \eta^{\#}$.
(c) $\eta^{\#}(M^c) = (\eta(M))^c$ and so $\eta^{\#}(M) = (\eta(M^c))^c$ and $\eta(M) = (\eta^{\#}(M^c))^c$.

Proof

(a) Firstly, let $\eta^{-1}(B) \subset M$ then using Lemma 3.1(a), it follows that $\eta^{\#}(\eta^{-1}(B)) \subset \eta^{\#}(M)$ and so $B \subset \eta^{\#}(\eta^{-1}(B)) \subset \eta^{\#}(M)$ using Lemma 3.1(c).
Conversely, let $B \subset \eta^{\#}(M)$ then $\eta^{-1}(B) \subset \eta^{-1}(\eta^{\#}(M))$. Therefore, by Lemma 3.1(b), it follows that $\eta^{-1}(B) \subset \eta^{-1}(\eta^{\#}(M)) \subset M$.

(b) Let $\eta : U \to V$ and $g : V \to Z$ be any two maps then $g \circ \eta : U \to Z$. Therefore, for each $z \in Z$, we have

$$\mu_{g^{\#} \circ \eta^{\#}(M)}(z) = \mu_{g^{\#}(\eta^{\#}(M))}(z)$$

$$= \begin{cases} \inf\limits_{v \in g^{-1}(z)} \mu_{\eta^{\#}(M)}(v) & g^{-1}(z) \neq \emptyset \\ 1 & g^{-1}(z) = \emptyset \end{cases}$$

$$= \begin{cases} \inf\limits_{v \in g^{-1}(z)} \begin{cases} \inf\limits_{u \in \eta^{-1}(v)} \mu_M(u), & \eta^{-1}(v) \neq \emptyset \\ 1, & \eta^{-1}(v) = \emptyset \end{cases}, & g^{-1}(z) \neq \emptyset \\ 1 & g^{-1}(z) = \emptyset \end{cases}$$

Now, three cases arise

(i) If $g^{-1}(z) \neq \emptyset$ and $\eta^{-1}(v) \neq \emptyset$ then $(g \circ \eta)^{-1}(z) \neq \emptyset$.

(ii) If $g^{-1}(z) \neq \emptyset$ and $\eta^{-1}(v) = \emptyset$ then $(g \circ \eta)^{-1}(z) = \emptyset$.

(iii) If $g^{-1}(z) = \emptyset$ then $(g \circ \eta)^{-1}(z) = \emptyset$.

It follows that

$$\mu_{g^{\#} \circ \eta^{\#}(M)}(z)$$

$$= \begin{cases} \inf\limits_{v \in g^{-1}(z)} \inf\limits_{u \in \eta^{-1}(v)} \mu_M(u) & g^{-1}(z) \neq \emptyset \text{ and } \eta^{-1}(v) \neq \emptyset \\ 1 & g^{-1}(z) \neq \emptyset \text{ and } \eta^{-1}(v) = \emptyset \\ 1 & g^{-1}(z) = \emptyset \end{cases}$$

$$= \begin{cases} \inf\limits_{u \in (g \circ \eta)^{-1}(z)} \mu_M(u) & (g \circ \eta)^{-1}(z) \neq \emptyset \\ 1 & (g \circ \eta)^{-1}(z) = \emptyset \end{cases}$$

$$= \mu_{(g \circ \eta)^{\#}(M)}(z)$$

Hence $(g \circ \eta)^{\#} = g^{\#} \circ \eta^{\#}$.

(c) For each $v \in V$, if $\eta^{-1}(v) \neq \emptyset$, then

$$\begin{aligned} \mu_{\eta^{\#}(M^c)}(v) &= \inf_{z \in \eta^{-1}(v)} \mu_{M^c}(z) \\ &= \inf_{z \in \eta^{-1}(v)} \{1 - \mu_M(z)\} \\ &= 1 - \sup_{z \in \eta^{-1}(v)} \mu_M(z) \\ &= 1 - \mu_{\eta(M)}(v) \\ &= \mu_{(\eta(M))^c}(v) \end{aligned}$$

and if $\eta^{-1}(v) = \emptyset$ *then*

$$\begin{aligned}\mu_{\eta^{\#}(M^c)}(v) &= 1 \\ \mu_{(\eta(M))^c}(v) &= 1 - \mu_{(\eta(M))}(v) \\ &= 1 - 0 \\ &= 1\end{aligned}$$

Hence $\eta^{\#}(M^c) = (\eta(M))^c$ *and so* $\eta(M) = (\eta^{\#}(M^c))^c$. *Replacing M by* M^c *we get,* $\eta^{\#}(M) = (\eta(M^c))^c$.

Definition 3.7 *A fuzzy subset M of U is called fuzzy saturated subset of U if* $M = \eta^{-1}(B)$ *for some fuzzy subset B of V i.e.* $\mu_M(u) = \mu_{\eta^{-1}(B)}(u)$ *for each* $u \in U$.

Remark 3.2 *For any map* $\eta : U \to V$, $M^{\#}$ *is fuzzy saturated for each fuzzy subset M of U, since,* $\eta^{-1}(\eta(M^{\#})) = \eta^{-1}(\eta(\eta^{-1}(\eta^{\#}(M)))) = \eta^{-1}(\eta^{\#}(M)) = M^{\#}$.

The following two Lemmas characterize fuzzy saturated sets in terms of image and # image of a fuzzy set.

Lemma 3.3 *A subset M of U is fuzzy saturated if and only if* $M = \eta^{-1}(\eta(M))$.

Proof *Let M be fuzzy saturated subset of U then* $M = \eta^{-1}(B)$ *for some fuzzy subset B of V. Therefore,* $\mu_M(u) = \mu_{\eta^{-1}(B)}(u)$ *for each* $u \in U$ *and so* $\mu_M(u_1) = \mu_M(u_2)$ *whenever* $\eta(u_1) = \eta(u_2)$. *Now, for each* $u \in U$

$$\begin{aligned}\mu_{\eta^{-1}(\eta(M))}(u) &= \mu_{\eta(M)}(\eta(u)) \\ &= \sup_{z \in \eta^{-1}(\eta(u))} \mu_M(z) \\ &= \mu_M(u),\end{aligned}$$

since for all $z \in \eta^{-1}(\eta(u))$, $\mu_M(z) = \mu_M(u)$. *Hence* $\eta^{-1}(\eta(M)) = M$.

Lemma 3.4 *For any map* $\eta : U \to V$ *and any fuzzy subset M in U, M is fuzzy saturated if and only if* $M = M^{\#}$.

Proof *Let M be any fuzzy saturated subset in U then*

$$\begin{aligned}\mu_{M^{\#}}(u) &= \mu_{\eta^{-1}(\eta^{\#}(M))}(u) \\ &= \mu_{\eta^{\#}(M)}(\eta(u)) \\ &= \inf_{z \in \eta^{-1}(\eta(u))} \mu_M(z) \\ &= \mu_M(u)\end{aligned}$$

Hence $M = M^{\#}$.
Conversely, let $M = M^{\#}$ *then by Remark 3.2, it follows that M is fuzzy saturated subset of U.*

The subsequent result gives a new characterization of surjective maps using # image of a fuzzy set.

Theorem 3.3 *Let* η *be a function from U to V where* $U = \{u\}$ *and* $V = \{v\}$ *be spaces of points then* η *is surjective if and only if* $\eta(M^{\#}) = \eta^{\#}(M)$.

Proof *For each $v \in V$, $\eta^{-1}(v) \neq \emptyset$, since η is surjective. Therefore,*

$$\begin{aligned}\mu_{\eta(M^{\#})}(v) &= \sup_{z \in \eta^{-1}(v)} \mu_{M^{\#}}(z) \\ &= \sup_{z \in \eta^{-1}(v)} \mu_{\eta^{-1}(\eta^{\#}(M))}(z) \\ &= \sup_{z \in \eta^{-1}(v)} \mu_{\eta^{\#}(M)}(\eta(z)) \\ &= \mu_{\eta^{\#}(M)}(v)\end{aligned}$$

Conversely, let $\eta(M^{\#}) = \eta^{\#}(M)$ then for each $v \in V$

$$\mu_{\eta(M^{\#})}(v) = \mu_{\eta^{\#}(M)}(v)$$

which is possible if and only if $\eta^{-1}(v) \neq \emptyset$ for each $v \in V$ i.e. if and only if η is surjective, since by definition
$\mu_{\eta(M^{\#})}(v) = 0$ and $\mu_{\eta^{\#}(M)}(v) = 1$, if $\eta^{-1}(v) = \emptyset$.
Equivalently, we can say that η is surjective if and only if $\eta^{\#}(\phi) = \phi$.

3.4 APPLICATIONS OF $\eta^{\#}$ TO FUZZY CONTINUITY

The subsequent theorem gives the characterization of fuzzy continuous maps in connection with the interior in fuzzy topology.

Theorem 3.4 *Let η be a function from U to V where $U = \{u\}$ and $V = \{v\}$ be spaces of points then η is fuzzy continuous iff for every fuzzy subset M of U, $(\eta^{\#}(M))^{o} \subset \eta^{\#}(M^{o})$.*

Proof *By Theorem 3.2, fuzzy continuity of η is equivalent to $\eta(\overline{M}) \subset \overline{\eta(M)}$, for each fuzzy subset M of U which is again equivalent to $[\eta^{\#}((\overline{M})^{c})]^{c} \subset \overline{(\eta^{\#}(M^{c}))^{c}}$ by part (c) of Lemma 3.2. But $\overline{[(\eta^{\#}(M^{c}))^{c}]} = [(\eta^{\#}(M^{c}))^{o}]^{c}$, since $\overline{(M^{c})} = (M^{o})^{c}$ and $[\eta^{\#}((\overline{M})^{c})]^{c} = [(\eta^{\#}((M^{c})^{o})]^{c}$ since $(\overline{M})^{c} = (M^{c})^{o}$ by Theorem 3.1. Therefore, $\eta(\overline{M}) \subset \overline{\eta(M)}$ is equivalent to $(\eta^{\#}(M^{c}))^{o} \subset \eta^{\#}((M^{c})^{o})$ which holds for arbitrary fuzzy subsets M of U. Hence η is fuzzy continuous if and only if for every fuzzy subset M of U, $(\eta^{\#}(M))^{o} \subset \eta^{\#}(M^{o})$.*

The next corollary characterizes fuzzy continuity for surjective mppings.

Corollary 3.1 *Let $\eta : U \to V$ be any surjective map. Then η is fuzzy continuous iff for every fuzzy subset M of U, $(\eta(M^{\#}))^{o} \subset \eta((M^{o})^{\#})$.*

Proof *The proof follows from Theorems 3.4 and 3.3.*

The following theorem gives equivalent conditions for a surjective fuzzy continuous map.

Theorem 3.5 *Let $\eta : U \to V$ be any surjective map. Then fuzzy continuity is equivalent to the following conditions:*

(a) η is fuzzy continuous.
(b) $M^{\#}$ is fuzzy open in U if $\eta(M^{\#})$ is fuzzy open in V.
(c) $M^{\#}$ is fuzzy open in U if $\eta^{\#}(M)$ is fuzzy open in V.
(d) for all fuzzy saturated set M in U, M is fuzzy closed in U if $\eta(M)$ is fuzzy closed in V.
(e) for all fuzzy saturated set M in U, M is fuzzy open in U if $\eta^{\#}(M)$ is fuzzy open in V.

Proof $(a) \Rightarrow (b)$: *Let* $\eta(M^{\#})$ *be fuzzy open in* V *then* η *is fuzzy continuous implies that* $\eta^{-1}\eta(M^{\#})$ *is fuzzy open in* U *and so* $\eta^{-1}(\eta(\eta^{-1}(\eta^{\#}(M))))$ *is fuzzy open in* U. *But* $\eta^{-1}\circ\eta\circ\eta^{-1} = \eta^{-1}$. *Therefore,* $\eta^{-1}(\eta^{\#}(M))$ *is fuzzy open in* U *and so* $M^{\#}$ *is fuzzy open in* U.

$(b) \Rightarrow (c)$: *Let* $\eta^{\#}(M)$ *be fuzzy open in* V *then* η *is surjective implies* $\eta(M^{\#})$ *is fuzzy open in* V *and so by (b),* $M^{\#}$ *is fuzzy open in* U.

$(c) \Rightarrow (d)$: *Let* $\eta(M)$ *be fuzzy closed in* V, *where* M *is a fuzzy saturated set in* U, *then* $(\eta(M))^{c}$ *is fuzzy open in* V, *and so by Lemma 3.2(c),* $\eta^{\#}(M^{c})$ *is fuzzy open in* V. *Therefore, by (c),* $(M^{c})^{\#}$ *is fuzzy open in* U, *and so* M^{c} *is fuzzy open in* U, *since* M *is fuzzy saturated implies* M^{c} *is fuzzy saturated. Hence* M *is fuzzy closed in* U *and so (d) holds.*

$(d) \Rightarrow (e)$: *Let* $\eta^{\#}(M)$ *be fuzzy open in* V *for fuzzy saturated subset* M *of* U. *Then* $(\eta^{\#}(M))^{c}$ *is fuzzy closed in* V *and so* $\eta(M^{c})$ *is fuzzy closed in* V *by Lemma 3.2(c). Therefore, (d) implies that* M^{c} *is fuzzy closed in* U *and so* M *is fuzzy open in* U *and so (e) is true.*

$(e) \Rightarrow (a)$: *Let* B *be any fuzzy open subset of* V *and let* $\eta^{-1}(B) = E$. *Then* $B = \eta^{\#}(\eta^{-1}(B)) = \eta^{\#}(M)$ *by Lemma 3.1(c). Therefore* $\eta^{\#}(M)$ *is fuzzy open in* V *and so by (e),* M *is fuzzy open in* U, *since* M *is fuzzy saturated subset of* U. *Hence* $\eta^{-1}(B)$ *is fuzzy open in* U *and so* η *is fuzzy continuous.*

The following theorem gives the characterization of fuzzy continuity using fuzzy saturated sets.

Theorem 3.6 *Let* $\eta : U \to V$ *be any surjective map. Then* η *is fuzzy continuous if and only if for any fuzzy saturated subset* M *of* U, M *is fuzzy open in* U *whenever* $\eta(M)$ *is fuzzy open in* V.

Proof *Let* η *be fuzzy continuous and* $\eta(M)$ *be fuzzy open in* V *for fuzzy saturated subset* M *of* U. *Then* $\eta^{-1}(\eta(M))$ *is fuzzy open in* U *and so* M *is fuzzy open in* U.
Contrarily, let B *be fuzzy open subset of* V *and* $\eta^{-1}(B) = E$. *Then* $\eta^{-1}(B)$ *is fuzzy saturated subset of* U *and* $\eta(\eta^{-1}(B)) = B$ *is fuzzy open in* V, *since* η *is surjective. Therefore,* $\eta^{-1}(B)$ *is fuzzy open in* U *and so* η *is fuzzy continuous.*

3.5 CONCLUSION

Topological spaces have not only applications in mathematics but in various other fields like physics, geographical systems, computers, chemistry, etc. But because of uncertainties and incomplete information about an element, it is difficult to apply the concept of topology to all real-life problems. To overcome this difficulty, fuzzy sets and hence fuzzy topological spaces were introduced which deal with the uncertainty of an object. In various real-life applications of fuzzy set theory, fuzzy functions play an important role, especially in fuzzy control and approximate reasoning. Motivated by the applications of fuzzy functions, we have introduced a new type of fuzzy function denoted by $\eta^{\#}$ and have discussed their properties. Further, it motivates to contribute to the theoretical study of fuzzy continuous functions. We can also study fuzzy-open and fuzzy-closed functions using $\eta^{\#}$ which will be our next target and to study fuzzy topological properties preserved by these maps.

Funding
No funding was received for conducting this study.

Compliance with ethical standards
Conflict of interest The authors declare that they have no conflict of interest.

Authors' contributions

Sandeep Kaur had the idea for the article, Nitakshi Goyal (Corresponding author) performed the literature search and both authors drafted and critically revised the work.

REFERENCES

1. M. Alimohammady, E. Ekici, S. Jafari and M. Roohi, *On fuzzy upper and lower contra-continuous multifunctions*, Iranian Journal of Fuzzy Systems, **8**(3), 2011, pp. 149-158.
2. A.V. Arkhangel'skii and V.I. Ponomarev, *Fundamentals of General Topology: Problems and Exercises*, Mathematics and Its Applications, Hindustan Publishing Corporation, New Delhi, India, 1984.
3. C.L. Chang, *Fuzzy Topological Spaces*, J. Math. Anal. Appl., **24**, 1968, pp. 182-189.
4. Pu Pao-Ming and Liu Ying-Ming, *Fuzzy Topology. I. Neighbourhood structure of a fuzzy point and Moore-Smith convergence*, J. Math. Anal. Appl., **76**, 1980, pp. 571-599.
5. Pu Pao-Ming and Liu Ying-Ming, *Fuzzy Topology. II. Product and Quotient Spaces*, J. Math. Anal. Appl., **77**, 1980, pp. 20-37.
6. R.H. Warren, *Continuity of mappings of fuzzy topological spaces*, Notices Amer. Math. Soc., **21**(3), 1974, p. A-451.
7. R.H. Warren, *Neighbourhoods, Bases and continuity in fuzzy topological spaces*, Rocky Mountain Journal of Mathematics, **8**(3), 1978, pp. 459-470.
8. L.A. Zadeh, *Fuzzy sets*, Inform. and Contr., **8**, 1965, pp. 338-353.
9. Z.G. Quan, *Fuzzy continuous function and its properties*, Fuzzy Sets and Systems, **43**(2), 1991, pp. 159-171.

4 Tri-Rotational Hypersurfaces Satisfying $\Delta \mathfrak{x} = \mathscr{M} \mathfrak{x}$ in $\mathbb{E}^7$

Erhan Guler
Bartin University, Faculty of Sciences, Department of Mathematics, Kutlubey Campus, Bartin, Turkey

Omer Kisi
Bartin University, Faculty of Sciences, Department of Mathematics, Kutlubey Campus, Bartin, Turkey

CONTENTS

We present the rotational hypersurface denoted by $\mathfrak{x}$ which has been constructed through triple rotation, having six parameters in the seven-dimensional Euclidean space $\mathbb{E}^7$. We proceed to disclose the fundamental forms, Gauss map, and shape operator of the aforementioned rotational hypersurface. Furthermore, by utilizing the Cayley–Hamilton theorem, we compute the curvatures of $\mathfrak{x}$. We also provide certain equations describing the relationships between the curvatures of $\mathfrak{x}$. Additionally, we unveil Laplace–Beltrami operator that $\Delta \mathfrak{x} = \mathscr{M} \mathfrak{x}$, where $\mathscr{M}$ represents a 7×7 matrix within $\mathbb{E}^7$.

4.1 INTRODUCTION

The concept of finite-dimensional submanifolds immersed in $\mathbb{E}^m$ or $\mathbb{E}^m_v$ utilizing a limited set of eigenfunctions derived from their Laplacian was originally proposed by Chen [4, 5, 6, 7]. Since then, this subject has been subjected to thorough investigation within the discipline.

A related Euclidean submanifold was investigated by Takahashi [20], being of 1-type (1-t) if and only if it is minimal or minimal in some hypersphere of $\mathbb{E}^m$. Hypersurfaces with constant scalar curvature were focused on by Cheng and Yau [10]. Takahashi's theorem in $\mathbb{E}^m$ was studied by Garay [14]. Research was conducted by Chen and Piccinni [9] on submanifolds in $\mathbb{E}^m$ that have a finite type Gauss map. Weyl pseudosymmetric hypersurfaces were explored by Arslan et al. [3]. Hypersurfaces with pointwise 1-t Gauss maps in $(n+1)$-space were introduced by Dursun [12]. The geometry of submanifolds was also introduced by Aminov [2]. Helical hypersurfaces were studied by Güler et al. [15]. The *III*-Laplacian of rotational hypersurfaces was determined by Güler et al. [16]. Four decades were dedicated by Chen et al. [8] to the investigation of submanifolds 1-t and

DOI: 10.1201/9781032634142-4

the 1-t Gauss map within the domain of space forms. Li and Güler [19] worked hypersurfaces of revolution family in pseudo-Euclidean space $\mathbb{E}^7_3$.

The research aims to investigate a rotational hypersurface, denoted as $\mathfrak{r}$, which is constructed through triple rotation in the seven-dimensional Euclidean space $\mathbb{E}^7$. The main objectives of this study are:

- to uncover the fundamental forms, Gauss map, and shape operator associated with the rotational hypersurface $\mathfrak{r}$,
- to employ the Cayley–Hamilton theorem for computing the curvatures and principal curvatures of $\mathfrak{r}$,
- to establish equations that describe the relationships between the mean curvature and Gauss–Kronecker curvature of $\mathfrak{r}$,
- to unveil the Laplace–Beltrami operator, denoted as $\Delta\mathfrak{r} = \mathcal{M}\mathfrak{r}$, where $\mathcal{M}$ represents a 7×7 matrix within $\mathbb{E}^7$.

In this work, our analysis primarily focuses on examining the tri-rotational hypersurface, which is dependent on six parameters and is situated in the seven-dimensional Euclidean space $\mathbb{E}^7$. Our research is structured as follows.

In Section 4.2, we provide a detailed explanation of the fundamental concepts of seven-dimensional Euclidean geometry, laying the groundwork for our subsequent analysis.

In Section 4.3, we present the curvature formulas specifically tailored for hypersurfaces in $\mathbb{E}^7$. This enables us to quantify and characterize the geometric properties of the tri-rotational hypersurface under investigation.

Section 4.4 is dedicated to offering a comprehensive definition of the tri-rotational hypersurface, outlining its construction method and key properties. By examining this particular type of hypersurface, we aim to gain deeper insights into its structural characteristics and behavior.

We introduce the Laplace–Beltrami operator for smooth functions in $\mathbb{E}^7$ in Section 4.5. Leveraging the insights and properties derived from the tri-rotational hypersurface discussed in the preceding section, we proceed to compute the Laplace–Beltrami operator and explore its implications within this context.

Finally, we conclude the work in Section 4.6.

Through these sections, we aim to provide a comprehensive analysis of the tri-rotational hypersurface, covering its geometry, curvature, definition, and the computation of the Laplace–Beltrami operator associated with it in the seven-dimensional Euclidean space.

4.2 PRELIMINARIES

In this work, we use the following notations, formulas, and equations:

- $\mathbb{E}^m$ denotes the Euclidean m-space where the coordinates are denoted by $(x_1, x_2, \ldots, x_m)$.
- The canonical Euclidean metric tensor on $\mathbb{E}^m$ is denoted by $\widetilde{g}$ and is given by $\widetilde{g} = \langle\ ,\ \rangle = \sum_{i=1}^{m} dx_i^2$.
- $\widetilde{M}$ is an m-dimensional Riemannian submanifold embedded in $\mathbb{E}^m$, while M represents its submanifold. The Levi-Civita connections associated with $\widetilde{M}$ are denoted as $\widetilde{\nabla}$ and ∇, respectively.
- We use letters X, Y, Z, and W to denote vector fields that are tangent to M, and ξ, η to denote vector fields that are normal to M.
- The Gauss formula:
$$\widetilde{\nabla}_X Y = \nabla_X Y + h(X, Y), \tag{4.1}$$
where h is the second fundamental form of M.

- The Weingarten formula:
$$\widetilde{\nabla}_X \xi = -A_\xi(X) + D_X \xi, \tag{4.2}$$
where A the shape operator and D is the normal connection of M.
- The shape operator A_ξ is a symmetric endomorphism of the tangent space T_pM at each point $p \in M$ for each $\xi \in T_p^\perp M$.
- There exists a relationship between the shape operator and the second fundamental form given by
$$\langle h(X,Y), \xi \rangle = \langle A_\xi X, Y \rangle.$$
- The Gauss equation:
$$\langle R(X,Y,)Z,W \rangle = \langle h(Y,Z), h(X,W) \rangle - \langle h(X,Z), h(Y,W) \rangle, \tag{4.3}$$
where R is the curvature tensor associated with the Levi–Civita connection ∇, and h is the second fundamental form of M.
- The Codazzi equation:
$$(\bar{\nabla}_X h)(Y,Z) = (\bar{\nabla}_Y h)(X,Z), \tag{4.4}$$
where $\bar{\nabla} h$ is the covariant derivative of h with respect to the Levi–Civita connection ∇, and X,Y,Z are tangent vector fields on M.
- The curvature tensor R^D associated with the normal connection D is not explicitly mentioned in the given equations.
- The covariant derivative of h, denoted as $(\bar{\nabla}_X h)(Y,Z)$, is defined by
$$(\bar{\nabla}_X h)(Y,Z) = D_X h(Y,Z) - h(\nabla_X Y, Z) - h(Y, \nabla_X Z).$$
where D is the normal connection of M.

Let M represent an oriented hypersurface situated in $\mathbb{E}^{n+1}$, equipped with its shape operator denoted as $\mathbf{S}$, and the position vector denoted as x. We examine an orthonormal frame field $\{e_1, e_2, \ldots, e_n\}$ comprising principal directions of M, which correspond to the principal curvatures k_i, with $i = 1, 2, \ldots, n$.

The first structural equation of Cartan can be expressed as

$$d\theta_i = \sum_{i=1}^{n} \theta_j \wedge \omega_{ij}, \quad i,j = 1,2,\ldots,n, \tag{4.5}$$

where ω_{ij} represents the connection forms coinciding with the chosen frame field. By the Codazzi equation (4.4), we can derive the following equations

$$e_i(k_j) = \omega_{ij}(e_j)(k_i - k_j), \tag{4.6}$$
$$\omega_{ij}(e_l)(k_i - k_j) = \omega_{il}(e_j)(k_i - k_l) \tag{4.7}$$

for different $i, j, l = 1, 2, \ldots, n$.

We let $s_j = \sigma_j(k_1, k_2, \ldots, k_n)$, were σ_j is the j-th elementary symmetric function defined by

$$\sigma_j(a_1, a_2, \ldots, a_n) = \sum_{1 \le i_1 < i_2 < \ldots < i_j \le n} a_{i_1} a_{i_2} \ldots a_{i_j}.$$

We introduce the following notation

$$r_i^j = \sigma_j(k_1, k_2, \ldots, k_{i-1}, k_{i+1}, k_{i+2}, \ldots, k_n).$$

Based on the provided definition, we observe that $r_i^0 = 1$ and $s_{n+1} = s_{n+2} = \cdots = 0$. s_k is commonly known as the k-th mean curvature of the oriented hypersurface M. The mean curvature $H = \frac{1}{n}s_1$ is also defined, and the Gauss–Kronecker curvature of M is $K = s_n$. If $s_j \equiv 0$, the hypersurface M is known as j-minimal.

In Euclidean $(n+1)$-space, getting the i-th curvature formulas $\mathscr{K}_i$ (See [1] and [17] for details.), where $i = 0, \ldots, n$, we have the following characteristic polynomial of $\mathbf{S}$:

$$\sum_{k=0}^{n} (-1)^k s_k \lambda^{n-k} = P_{\mathbf{S}}(\lambda) = \det(\mathbf{S} - \lambda \mathscr{I}_n) = 0. \tag{4.8}$$

Here $i = 0, \ldots, n$, $\mathscr{I}_n$ indicates the identity matrix. Hence, we reveal the curvature formulas as $\binom{n}{i}\mathscr{K}_i = s_i$. Further details regarding this relationship can be found in the reference [5].

Let $\mathfrak{x} = \mathfrak{x}(u, v, w, s, p, q)$ be an immersion from $M^6 \subset \mathbb{E}^6$ to $\mathbb{E}^7$. A sixtuple vector product of $\overrightarrow{x^1} = (x_1^1, x_2^1, \ldots, x_7^1), \ldots, \overrightarrow{x^6} = (x_1^6, x_2^6, \ldots, x_7^6)$ of $\mathbb{E}^7$ is defined by

$$\overrightarrow{x^1} \times \overrightarrow{x^2} \times \cdots \times \overrightarrow{x^6} = \det \begin{pmatrix} e_1 & e_2 & \cdots & e_7 \\ x_1^1 & x_2^1 & \cdots & x_7^1 \\ \vdots & \vdots & \ddots & \vdots \\ x_1^6 & x_2^6 & \cdots & x_7^6 \end{pmatrix}.$$

For a hypersurface $\mathfrak{x}$ in 7-space, $(\mathbf{g}_{ij})_{6\times6}$, $(\mathbf{h}_{ij})_{6\times6}$ describes the first and the second fundamental form matrices, respectively, and $g_{ij} = \langle \mathfrak{x}_i, \mathfrak{x}_j \rangle$, $h_{ij} = \langle \mathfrak{x}_{ij}, \mathbb{G} \rangle$, $i, j = 1, 2, \ldots, 6$, $\mathfrak{x}_u = \frac{\partial \mathfrak{x}}{\partial u}$ when $i = 1$, $\mathfrak{x}_{uv} = \frac{\partial^2 \mathfrak{x}}{\partial u \partial v}$ when $i = 1$ and $j = 2$. Here,

$$\mathbb{G} = \frac{\mathfrak{x}_u \times \mathfrak{x}_v \times \mathfrak{x}_w \times \mathfrak{x}_s \times \mathfrak{x}_p \times \mathfrak{x}_q}{\left\| \mathfrak{x}_u \times \mathfrak{x}_v \times \mathfrak{x}_w \times \mathfrak{x}_s \times \mathfrak{x}_p \times \mathfrak{x}_q \right\|} \tag{4.9}$$

determines the Gauss map of the hypersurface $\mathfrak{x}$.

4.3 CURVATURES

In this section, we provide the curvature formulas for an arbitrary hypersurface having six parameters in the seven-dimensional Euclidean space $\mathbb{E}^7$.

Theorem 4.1 *A hypersurface $\mathfrak{x}$ in $\mathbb{E}^7$ has the following curvature formulas, $\mathscr{K}_0 = 1$ by definition,*

$$6\mathscr{K}_1 = -\frac{\mathfrak{b}}{\mathfrak{a}},\ 15\mathscr{K}_2 = \frac{\mathfrak{c}}{\mathfrak{a}},\ 20\mathscr{K}_3 = -\frac{\mathfrak{d}}{\mathfrak{a}},\ 15\mathscr{K}_4 = \frac{\mathfrak{e}}{\mathfrak{a}},\ 6\mathscr{K}_5 = -\frac{\mathfrak{f}}{\mathfrak{a}},\ \mathscr{K}_6 = \frac{\mathfrak{g}}{\mathfrak{a}}, \tag{4.10}$$

where $P_{\mathbf{S}}(\lambda) = \mathfrak{a}\lambda^6 + \mathfrak{b}\lambda^5 + \mathfrak{c}\lambda^4 + \mathfrak{d}\lambda^3 + \mathfrak{e}\lambda^2 + \mathfrak{f}\lambda + \mathfrak{g} = 0$ denotes the characteristic polynomial of the shape operator matrix $\mathbf{S}$, $\mathfrak{a} = \det(\mathbf{g}_{ij})$, $\mathfrak{g} = \det(\mathbf{h}_{ij})$, and $(\mathbf{g}_{ij})$ and $(\mathbf{h}_{ij})$ are the first, and the second fundamental form matrices, respectively.

Proof *The solution matrix $(\mathbf{g}_{ij})^{-1} \cdot (\mathbf{h}_{ij})$ supplies the shape operator matrix $\mathbf{S}$ of hypersurface $\mathfrak{x}$ in 7-space. In $\mathbb{E}^7$, computing the i-th mean curvature formula $\mathscr{K}_i$, where $i = 0, 1, \ldots, 6$, we reveal the characteristic polynomial $P_{\mathbf{S}}(\lambda) = \det(\mathbf{S} - \lambda \mathscr{I}_6) = \mathfrak{a}\lambda^6 + \mathfrak{b}\lambda^5 + \mathfrak{c}\lambda^4 + \mathfrak{d}\lambda^3 + \mathfrak{e}\lambda^2 + \mathfrak{f}\lambda + \mathfrak{g} = 0$*

of **S**. *Then, we find the i-th curvature folmulas in 7-space. We obtain the following*

$$\binom{6}{0}\mathscr{K}_0 = 1,$$
$$\binom{6}{1}\mathscr{K}_1 = \sum_{i=1}^{6} k_i = -\frac{\mathfrak{b}}{\mathfrak{a}},$$
$$\binom{6}{2}\mathscr{K}_2 = \sum_{1=i_1<i_2}^{6} k_{i_1}k_{i_2} = \frac{\mathfrak{c}}{\mathfrak{a}},$$
$$\binom{6}{3}\mathscr{K}_3 = \sum_{1=i_1<i_2<i_3}^{6} k_{i_1}k_{i_2}k_{i_3} = -\frac{\mathfrak{d}}{\mathfrak{a}},$$
$$\binom{6}{4}\mathscr{K}_4 = \sum_{1=i_1<i_2<i_3<i_4}^{6} k_{i_1}k_{i_2}k_{i_3}k_{i_4} = \frac{\mathfrak{e}}{\mathfrak{a}},$$
$$\binom{6}{5}\mathscr{K}_5 = \sum_{1=i_1<i_2<i_3<i_4<i_5}^{6} k_{i_1}k_{i_2}k_{i_3}k_{i_4}k_{i_5} = -\frac{\mathfrak{f}}{\mathfrak{a}},$$
$$\binom{6}{6}\mathscr{K}_6 = \prod_{i=1}^{6} k_i = \frac{\mathfrak{g}}{\mathfrak{a}},$$

Theorem 4.2 *A hypersurface* $\mathfrak{x} = \mathfrak{x}(u,v,w,s,p,q)$ *in* $\mathbb{E}^7$ *has the following relation*

$$\mathscr{K}_0 VII - 6\mathscr{K}_1 VI + 15\mathscr{K}_2 V - 20\mathscr{K}_3 IV + 15\mathscr{K}_4 III - 6\mathscr{K}_5 II + \mathscr{K}_6 I = 0,$$

where $I, II, \ldots, VII$ *determines the fundamental form matrices having order* 6×6 *of the hypersurface.*

Proof *Regarding* $n = 6$ *in* (4.8), *it works.*

4.4 TRI-ROTATIONAL HYPERSURFACE IN EUCLIDEAN 7-SPACE

In this section, we provide a definition for the tri-rotational hypersurface and explore its differential geometric properties within the Euclidean 7-space $\mathbb{E}^7$. The study of rotational hypersurfaces in Riemannian space forms is discussed in detail in [11].

A rotational hypersurface M in Euclidean $(n+1)$-space is constructed by revolving a surface γ around a rotating axis ℓ (see [11] for reference).

To construct the rotational hypersurface we start with the generating hypersurface given by $\gamma = \gamma(u,v,w) = (f,0,g,0,h,0,k)$, and apply the rotation matrix $\mathfrak{R} = \operatorname{diag}(\mathscr{R}_s, \mathscr{R}_p, \mathscr{R}_q, 1)$ with the elements given by $\mathscr{R}_j = \begin{pmatrix} \cos j & -\sin j \\ \sin j & \cos j \end{pmatrix}$, $j = s,p,q$. Then, we state the rotational hypersurface $\mathfrak{x} = \mathfrak{R} \cdot \gamma^T$ when γ rotates about axis $\ell = (0,0,0,0,0,0,1)$, and present the following.

Definition 4.1 *A rotational hypersurface is an immersion* $\mathfrak{x} : M^6 \subset \mathbb{E}^6 \longrightarrow \mathbb{E}^7$ *having triple rotation, with rotating axis* $(0,0,0,0,0,0,1)$, *defined by*

$$\mathfrak{x}(u,v,w,s,p,q) = (f\cos s, f\sin s, g\cos p, g\sin p, h\cos q, h\sin q, k), \tag{4.11}$$

where f,g,h,k *denotes the differentiable functions, and depend on* $u,v,w \in \mathbb{R}$, $0 \le s,p,q < 2\pi$.

Throughout the work, we referred to hypersurface $\mathfrak{x}$ determined by Eq. (4.11) as a *tri-rotational hypersurface*.

Considering the first derivatives of (4.11) w.r.t. u, v, w, s, p, q, respectively, we find the first quantities of (4.11):

$$(\mathbf{g}_{ij}) = \operatorname{diag}\left((\mathbf{g}_{kl})_{3\times 3},\quad \mathbf{g}_{44},\quad \mathbf{g}_{55},\quad \mathbf{g}_{66}\right), \tag{4.12}$$

where

$$\begin{aligned}
\mathbf{g}_{11} &= f_u^2 + g_u^2 + h_u^2 + k_u^2,\\
\mathbf{g}_{12} &= f_u f_v + g_u g_v + h_u h_v + k_u k_v = \mathbf{g}_{21},\\
\mathbf{g}_{13} &= f_u f_w + g_u g_w + h_u h_w + k_u k_w = \mathbf{g}_{31},\\
\mathbf{g}_{22} &= f_v^2 + g_v^2 + h_v^2 + k_v^2,\\
\mathbf{g}_{23} &= f_v f_w + g_v g_w + h_v h_w + k_v k_w = \mathbf{g}_{32},\\
\mathbf{g}_{33} &= f_w^2 + g_w^2 + h_w^2 + k_w^2,\\
\mathbf{g}_{44} &= f^2,\\
\mathbf{g}_{55} &= g^2,\\
\mathbf{g}_{66} &= h^2,
\end{aligned}$$

and $f_u = \frac{\partial f}{\partial u}$, $f_v = \frac{\partial f}{\partial v}$, $f_u^2 = \frac{\partial^2 f}{\partial u^2}$, etc.. Here,

$$\begin{aligned}
\hat{\mathbf{g}} = \det(\mathbf{g}_{ij}) &= \left(\mathbf{g}_{11}\mathbf{g}_{22}\mathbf{g}_{33} + 2\mathbf{g}_{12}\mathbf{g}_{13}\mathbf{g}_{23} - \mathbf{g}_{11}\mathbf{g}_{23}^2 - \mathbf{g}_{13}^2\mathbf{g}_{22} - \mathbf{g}_{12}^2\mathbf{g}_{33}\right)\mathbf{g}_{44}\mathbf{g}_{55}\mathbf{g}_{66}\\
&= f^2 g^2 h^2 W,
\end{aligned}$$

and

$$\begin{aligned}
W = 2(&f_u f_v g_u g_w h_v h_w + f_u f_v g_v g_w h_u h_w + f_u f_w g_u g_v h_v h_w + f_u f_w g_w g_v h_u h_v + f_v f_w g_u g_v h_u h_w\\
&+ f_v g_u f_w h_u g_w h_v + f_u f_v g_u g_w k_v k_w + f_u f_v g_v g_w k_u k_w + f_u g_u f_w g_v k_v k_w + f_u f_w g_v g_w k_u k_v\\
&+ f_v g_u f_w g_v k_u k_w + f_v g_u f_w g_w k_u k_v + f_u f_v h_u h_w k_v k_w + f_u f_v h_v h_w k_u k_w + f_u f_w h_u h_v k_v k_w\\
&+ f_u f_w h_v h_w k_u k_v + f_v f_w h_u h_v k_u k_w + f_v f_w h_u h_w k_u k_v + g_u g_v h_u h_w k_v k_w + g_u g_v h_v h_w k_u k_w\\
&+ g_u h_u g_w h_v k_v k_w + g_u g_w h_v h_w k_u k_v + g_v h_u g_w h_v k_u k_w + g_v h_u g_w h_w k_u k_v - f_u f_v g_u g_v h_w^2\\
&- f_u f_v h_u g_w^2 h_v - f_u g_u f_w g_w h_v^2 - f_u f_w g_v^2 h_u h_w - f_v f_w g_v h_u^2 g_w - g_u f_w^2 g_v h_u h_v - f_u f_v g_u g_v k_w^2\\
&- f_v g_u^2 f_w h_v h_w - f_v^2 g_u h_u g_w h_w - f_u g_u f_w g_w k_v^2 - f_u^2 g_v g_w h_v h_w - f_u f_v h_u h_v k_w^2 - f_v f_w g_v g_w k_u^2\\
&- f_u f_w h_u h_w k_v^2 - f_u f_v g_w^2 k_u k_v - f_v f_w h_v h_w k_u^2 - g_u g_v h_u h_v k_w^2 - f_u f_v h_w^2 k_u k_v - f_u f_w g_v^2 k_u k_w\\
&- g_u h_u g_w h_w k_v^2 - g_u f_w^2 g_v k_u k_v - f_u f_w h_v^2 k_u k_w - f_v g_u^2 f_w k_v k_w - g_v g_w h_v h_w k_u^2 - f_v^2 g_u g_w k_u k_w\\
&- f_v f_w h_u^2 k_v k_w - g_u g_v h_w^2 k_u k_v - f_u^2 g_v g_w k_v k_w - f_w^2 h_u h_v k_u k_v - g_u g_w h_v^2 k_u k_w - h_u g_w^2 h_v k_u k_v\\
&- f_v^2 h_u h_w k_u k_w - g_v h_u^2 g_w k_v k_w - f_u^2 h_v h_w k_v k_w - g_v^2 h_u h_w k_u k_w - g_u^2 h_v h_w k_v k_w)\\
&+ f_u^2 g_v^2 h_w^2 + f_u^2 g_w^2 h_v^2 + f_v^2 g_u^2 h_w^2 + f_v^2 g_w^2 h_u^2 + f_w^2 g_u^2 h_v^2 + f_w^2 g_v^2 h_u^2\\
&+ f_u^2 g_v^2 k_w^2 + f_u^2 g_w^2 k_v^2 + f_v^2 g_u^2 k_w^2 + f_v^2 g_w^2 k_u^2 + f_w^2 g_u^2 k_v^2 + f_w^2 g_v^2 k_u^2\\
&+ f_u^2 h_v^2 k_w^2 + f_u^2 h_w^2 k_v^2 + f_v^2 h_u^2 k_w^2 + f_v^2 h_w^2 k_u^2 + f_w^2 h_u^2 k_v^2 + f_w^2 h_v^2 k_u^2\\
&+ g_u^2 h_v^2 k_w^2 + g_u^2 h_w^2 k_v^2 + g_v^2 h_u^2 k_w^2 + g_v^2 h_w^2 k_u^2 + g_w^2 h_u^2 k_v^2 + g_w^2 h_v^2 k_u^2.
\end{aligned}$$

Using (4.9), we obtain the following Gauss map of the tri-rotational hypersurface (4.11):

$$\mathbb{G} = \frac{1}{W^{1/2}}\left(G_1\cos s, G_1\sin s, G_2\cos p, G_2\sin p, G_3\cos q, G_3\sin q, G_4\right), \tag{4.13}$$

where

$$\begin{aligned}
G_1 &= (g_v h_w - g_w h_v)k_u + (g_w h_u - g_u h_w)k_v + (g_u h_v - g_v h_u)k_w,\\
G_2 &= (f_v h_w - f_w h_v)k_u + (f_w h_u - f_u h_w)k_v + (f_u h_v - f_v h_u)k_w,\\
G_3 &= (f_v g_w - f_w g_v)k_u + (f_w g_u - f_u g_w)k_v + (f_u g_v - f_v g_u)k_w,\\
G_4 &= -(f_v g_w - f_w g_v)h_u - (f_w g_u - f_u g_w)h_v - (f_u g_v - f_v g_u)h_w
\end{aligned}$$

Taking the second derivatives w.r.t. u, v, w, s, p, q, and by the Gauss map (12.9) of (4.11), we reveal the second quantities

$$(\mathbf{h}_{ij}) = \text{diag}\left((\mathbf{h}_{kl})_{3\times3}, \quad \mathbf{h}_{44}, \quad \mathbf{h}_{55}, \quad \mathbf{h}_{66} \right), \tag{4.14}$$

where

$$\begin{aligned}
\mathbf{h}_{11} &= \frac{G_1 f_{uu} + G_2 g_{uu} + G_3 h_{uu} + G_4 k_{uu}}{W^{1/2}}, \\
\mathbf{h}_{12} &= \frac{G_1 f_{uv} + G_2 g_{uv} + G_3 h_{uv} + G_4 k_{uv}}{W^{1/2}} = \mathbf{h}_{21}, \\
\mathbf{h}_{13} &= \frac{G_1 f_{uw} + G_2 g_{uw} + G_3 h_{uw} + G_4 k_{uw}}{W^{1/2}} = \mathbf{h}_{31}, \\
\mathbf{h}_{22} &= \frac{G_1 f_{vv} + G_2 g_{vv} + G_3 h_{vv} + G_4 k_{vv}}{W^{1/2}}, \\
\mathbf{h}_{23} &= \frac{G_1 f_{vw} + G_2 g_{vw} + G_3 h_{vw} + G_4 k_{vw}}{W^{1/2}} = \mathbf{h}_{32}, \\
\mathbf{h}_{33} &= \frac{G_1 f_{ww} + G_2 g_{ww} + G_3 h_{ww} + G_4 k_{ww}}{W^{1/2}}, \\
\mathbf{h}_{44} &= \frac{G_1 f}{W^{1/2}}, \ \mathbf{h}_{55} = \frac{G_2 g}{W^{1/2}}, \ \mathbf{h}_{66} = \frac{G_3 h}{W^{1/2}}, \\
W &= (G_1)^2 + (G_2)^2 + (G_3)^2 + (G_4)^2.
\end{aligned}$$

and $f_{uu} = \frac{\partial^2 f}{\partial u^2}$, $f_{uv} = \frac{\partial^2 f}{\partial u \partial v}$, etc. By using (4.12) and (4.14), we compute the shape operator matrix of (4.11):

$$\mathbf{S} = \text{diag}\left((s_{kl})_{3\times3}, \quad s_{44}, \quad s_{55}, \quad s_{66} \right)$$

with the following components

$$\begin{aligned}
s_{11} &= \left(\left(\mathbf{g}_{22}\mathbf{g}_{33} - \mathbf{g}_{23}^2\right)\mathbf{h}_{11} + \left(\mathbf{g}_{13}\mathbf{g}_{23} - \mathbf{g}_{12}\mathbf{g}_{33}\right)\mathbf{h}_{12} + \left(\mathbf{g}_{12}\mathbf{g}_{23} - \mathbf{g}_{13}\mathbf{g}_{22}\right)\mathbf{h}_{13} \right)/W, \\
s_{12} &= \left(\left(\mathbf{g}_{22}\mathbf{g}_{33} - \mathbf{g}_{23}^2\right)\mathbf{h}_{12} + \left(\mathbf{g}_{13}\mathbf{g}_{23} - \mathbf{g}_{12}\mathbf{g}_{33}\right)\mathbf{h}_{22} + \left(\mathbf{g}_{12}\mathbf{g}_{23} - \mathbf{g}_{13}\mathbf{g}_{22}\right)\mathbf{h}_{23} \right)/W, \\
s_{13} &= \left(\left(\mathbf{g}_{22}\mathbf{g}_{33} - \mathbf{g}_{23}^2\right)\mathbf{h}_{13} + \left(\mathbf{g}_{13}\mathbf{g}_{23} - \mathbf{g}_{12}\mathbf{g}_{33}\right)\mathbf{h}_{23} + \left(\mathbf{g}_{12}\mathbf{g}_{23} - \mathbf{g}_{13}\mathbf{g}_{22}\right)\mathbf{h}_{33} \right)/W, \\
s_{21} &= \left(\left(\mathbf{g}_{13}\mathbf{g}_{23} - \mathbf{g}_{12}\mathbf{g}_{33}\right)\mathbf{h}_{11} + \left(\mathbf{g}_{11}\mathbf{g}_{33} - \mathbf{g}_{13}^2\right)\mathbf{h}_{12} + \left(\mathbf{g}_{12}\mathbf{g}_{13} - \mathbf{g}_{11}\mathbf{g}_{23}\right)\mathbf{h}_{13} \right)/W, \\
s_{22} &= \left(\left(\mathbf{g}_{13}\mathbf{g}_{23} - \mathbf{g}_{12}\mathbf{g}_{33}\right)\mathbf{h}_{12} + \left(\mathbf{g}_{11}\mathbf{g}_{33} - \mathbf{g}_{13}^2\right)\mathbf{h}_{22} + \left(\mathbf{g}_{12}\mathbf{g}_{13} - \mathbf{g}_{11}\mathbf{g}_{23}\right)\mathbf{h}_{23} \right)/W, \\
s_{23} &= \left(\left(\mathbf{g}_{13}\mathbf{g}_{23} - \mathbf{g}_{12}\mathbf{g}_{33}\right)\mathbf{h}_{13} + \left(\mathbf{g}_{11}\mathbf{g}_{33} - \mathbf{g}_{13}^2\right)\mathbf{h}_{23} + \left(\mathbf{g}_{12}\mathbf{g}_{13} - \mathbf{g}_{11}\mathbf{g}_{23}\right)\mathbf{h}_{33} \right)/W, \\
s_{31} &= \left(\left(\mathbf{g}_{12}\mathbf{g}_{23} - \mathbf{g}_{13}\mathbf{g}_{22}\right)\mathbf{h}_{11} + \left(\mathbf{g}_{12}\mathbf{g}_{13} - \mathbf{g}_{11}\mathbf{g}_{23}\right)\mathbf{h}_{12} + \left(\mathbf{g}_{11}\mathbf{g}_{22} - \mathbf{g}_{12}^2\right)\mathbf{h}_{13} \right)/W, \\
s_{32} &= \left(\left(\mathbf{g}_{12}\mathbf{g}_{23} - \mathbf{g}_{13}\mathbf{g}_{22}\right)\mathbf{h}_{12} + \left(\mathbf{g}_{12}\mathbf{g}_{13} - \mathbf{g}_{11}\mathbf{g}_{23}\right)\mathbf{h}_{22} + \left(\mathbf{g}_{11}\mathbf{g}_{22} - \mathbf{g}_{12}^2\right)\mathbf{h}_{23} \right)/W, \\
s_{33} &= \left(\left(\mathbf{g}_{12}\mathbf{g}_{23} - \mathbf{g}_{13}\mathbf{g}_{22}\right)\mathbf{h}_{13} + \left(\mathbf{g}_{12}\mathbf{g}_{13} - \mathbf{g}_{11}\mathbf{g}_{23}\right)\mathbf{h}_{23} + \left(\mathbf{g}_{11}\mathbf{g}_{22} - \mathbf{g}_{12}^2\right)\mathbf{h}_{33} \right)/W, \\
s_{44} &= \frac{\mathbf{h}_{44}}{\mathbf{g}_{44}}, \quad s_{55} = \frac{\mathbf{h}_{55}}{\mathbf{g}_{55}}, \quad s_{66} = \frac{\mathbf{h}_{66}}{\mathbf{g}_{66}}.
\end{aligned}$$

Finally, using (4.10), with (4.12), (4.14), respectively, we find the curvatures of the tri-rotational hypersurface (4.11) as follows.

Theorem 4.3 *Let $\mathfrak{x}$ be a tri-rotational hypersurface given by Eq.* (4.11). *Then, $\mathfrak{x}$ has the principal curvatures*

$$
\begin{aligned}
k_1 &= \frac{1}{3}\Theta + \Omega^{1/2} + \frac{1}{9}\Omega^{-1/2}\Gamma, \\
k_2 &= \frac{1}{3}\Theta - \frac{1}{2}\Omega^{1/2} - \frac{1}{18}\left(\Omega^{-1/2} - \sqrt{-3}\right)\Gamma, \\
k_3 &= \frac{1}{3}\Theta - \frac{1}{2}\Omega^{1/2} - \frac{1}{18}\left(\Omega^{-1/2} + \sqrt{-3}\right)\Gamma, \\
k_4 &= s_{44}, \\
k_5 &= s_{55}, \\
k_6 &= s_{66},
\end{aligned}
$$

where $\Theta = s_{11} + s_{22} + s_{33}$, $\Gamma = s_{11}^2 + s_{22}^2 + s_{33}^2 + 3\left(s_{12}^2 + s_{13}^2 + s_{23}^2\right) - \left(s_{11}s_{33} + s_{22}s_{33} + s_{11}s_{22}\right)$, $\Omega = \Upsilon + \Lambda^{1/2}$,

$$
\begin{aligned}
\Upsilon = \tfrac{1}{54}[2&\left(s_{11}^3 + s_{22}^3 + s_{33}^3\right) - 3\left(s_{11}s_{22}^2 + s_{11}^2 s_{22} + s_{11}s_{33}^2 + s_{11}^2 s_{33} + s_{22}s_{33}^2 + s_{22}^2 s_{33}\right) \\
&+ 9\left(s_{11}s_{12}^2 + s_{11}s_{13}^2 + s_{22}s_{23}^2 + s_{12}^2 s_{22} + s_{13}^2 s_{33} + s_{23}^2 s_{33}\right) \\
&- 18\left(s_{11}s_{23}^2 + s_{13}^2 s_{22} + s_{12}^2 s_{33}\right) + 54 s_{12}s_{13}s_{23} + 12 s_{11}s_{22}s_{33}],
\end{aligned}
$$

and

$$
\begin{aligned}
\Lambda = \tfrac{1}{108}[&-4\left(s_{12}^6 + s_{13}^6 + s_{23}^6 + s_{11}^4 s_{23}^2 + s_{13}^2 s_{22}^4 + s_{12}^2 s_{33}^4\right) \\
&- 12\left(s_{12}^2 s_{13}^4 + s_{13}^2 s_{23}^4 + s_{12}^4 s_{13}^2 + s_{12}^2 s_{23}^4\right) \\
&+ \left(8 s_{11}^2 s_{23}^4 + s_{13}^4 s_{22}^2 + s_{12}^4 s_{33}^2\right) \\
&+ 10(\begin{pmatrix} s_{12}^4 s_{22}s_{33} + s_{13}^4 s_{22}s_{33} + s_{11}s_{12}^4 s_{22} \\ + s_{11}s_{13}^4 s_{33} + s_{22}s_{23}^4 s_{33} + s_{11}s_{23}^4 s_{33} \end{pmatrix} \\
&- 8\left(s_{11}s_{13}^4 s_{22} + s_{11}s_{22}s_{23}^4 + s_{11}s_{12}^4 s_{33}\right) \\
&+ 2\begin{pmatrix} s_{11}s_{22}s_{33}^4 + s_{11}s_{22}^4 s_{33} + s_{11}^4 s_{22}s_{33} \\ + s_{11}s_{12}^2 s_{22}^3 + s_{11}^3 s_{12}^2 s_{22} + s_{11}^3 s_{13}^2 s_{33} \\ + s_{11}s_{13}^2 s_{33}^3 + s_{22}s_{23}^2 s_{33}^3 + s_{22}^3 s_{23}^2 s_{33} \\ + s_{11}s_{12}^2 s_{13}^2 s_{22} + s_{11}s_{12}^2 s_{13}^2 s_{33} + s_{11}s_{13}^2 s_{23}^2 s_{33} \\ + s_{12}^2 s_{22}s_{23}^2 s_{33} + s_{13}^2 s_{22}s_{23}^2 s_{33} + s_{11}s_{12}^2 s_{22}s_{23}^2 \end{pmatrix} \\
&- +2\left(s_{11}^3 s_{22}^3 + s_{22}^3 s_{33}^3 + s_{11}^3 s_{33}^3\right) \\
&- 2\begin{pmatrix} s_{11}^3 s_{13}^2 s_{22} + s_{11}^3 s_{22}^2 s_{33} + s_{11}^3 s_{12}^2 s_{33} + s_{11}s_{22}^3 s_{33}^2 \\ + s_{11}^2 s_{22}s_{33}^3 + s_{11}^2 s_{22}^3 s_{33} + s_{12}^2 s_{22}^3 s_{33} + s_{11}^3 s_{22}s_{33}^2 \\ + s_{11}s_{23}^2 s_{33}^3 + s_{11}s_{22}^3 s_{23}^2 + s_{13}^2 s_{22}s_{33}^3 + s_{11}s_{22}^2 s_{33}^3 \end{pmatrix} \\
&- 72\left(s_{11}s_{12}s_{13}s_{23}^3 + s_{12}s_{13}^3 s_{22}s_{23} + s_{12}^3 s_{13}s_{23}s_{33}\right) \\
&+ 36\begin{pmatrix} s_{11}s_{12}s_{13}^3 s_{23} + s_{11}s_{12}^3 s_{13}s_{23} + s_{12}s_{13}s_{22}s_{23}^3 \\ + s_{12}^3 s_{13}s_{22}s_{23} + s_{12}s_{13}s_{23}^3 s_{33} + s_{12}s_{13}^3 s_{23}s_{33} \end{pmatrix}
\end{aligned}
$$

$$
\begin{aligned}
&+8\left(\begin{array}{c} s_{11}^3 s_{12}s_{13}s_{23} + s_{12}s_{13}s_{22}^3 s_{23} + s_{12}s_{13}s_{23}s_{33}^3 \\ +s_{11}s_{13}^2 s_{22}^3 + s_{11}s_{12}^2 s_{33}^3 + s_{11}^3 s_{22}s_{23}^2 \\ +s_{11}^3 s_{23}^2 s_{33} + s_{12}^2 s_{22}s_{33}^3 + s_{13}^2 s_{22}^3 s_{33} \end{array}\right) \\
&-12\left(\begin{array}{c} s_{11}s_{12}s_{13}s_{22}^2 s_{23} + s_{11}^2 s_{12}s_{13}s_{22}s_{23} + s_{11}^2 s_{12}s_{13}s_{23}s_{33} \\ +s_{11}s_{12}s_{13}s_{23}s_{33}^2 + s_{12}s_{13}s_{22}s_{23}s_{33}^2 + s_{12}s_{13}s_{22}^2 s_{23}s_{33} \end{array}\right) \\
&-\left(s_{11}^2 s_{12}^4 + s_{11}^2 s_{13}^4 + s_{11}^2 s_{22}^4 + s_{11}^4 s_{22}^2 + s_{12}^4 s_{22}^2 + s_{11}^2 s_{33}^4\right) \\
&-\left(s_{11}^4 s_{33}^2 + s_{22}^2 s_{23}^4 + s_{22}^2 s_{33}^4 + s_{13}^4 s_{33}^2 + s_{22}^4 s_{33}^2 + s_{23}^4 s_{33}^2\right) \\
&-12\left(s_{12}^4 s_{23}^2 + s_{13}^4 s_{23}^2\right) \\
&+10\left(\begin{array}{c} s_{11}s_{12}^2 s_{22}^2 s_{33} + s_{11}^2 s_{12}^2 s_{22}s_{33} + s_{11}s_{13}^2 s_{22}s_{33}^2 \\ +s_{11}^2 s_{13}^2 s_{22}s_{33} + s_{11}s_{22}s_{23}^2 s_{33}^2 + s_{11}s_{22}^2 s_{23}^2 s_{33} \end{array}\right) \\
&-20\left(s_{11}s_{12}^2 s_{22}s_{33}^2 + s_{11}s_{13}^2 s_{22}^2 s_{33} + s_{11}^2 s_{22}s_{23}^2 s_{33}\right) \\
&+38\left(s_{11}s_{13}^2 s_{22}s_{23}^2 + s_{11}s_{12}^2 s_{23}^2 s_{33} + s_{12}^2 s_{13}^2 s_{22}s_{33}\right) \\
&-8\left(s_{11}^2 s_{12}^2 s_{22}^2 + s_{11}^2 s_{13}^2 s_{33}^2 + s_{22}^2 s_{23}^2 s_{33}^2\right) + 6s_{11}^2 s_{22}^2 s_{33}^2 \\
&-2\left(\begin{array}{c} s_{11}^2 s_{12}^2 s_{13}^2 + s_{11}^2 s_{13}^2 s_{22}^2 + s_{11}^2 s_{12}^2 s_{33}^2 \\ +s_{11}^2 s_{22}^2 s_{23}^2 + s_{13}^2 s_{22}^2 s_{33}^2 + s_{13}^2 s_{23}^2 s_{33}^2 \\ +s_{12}^2 s_{22}^2 s_{23}^2 + s_{11}^2 s_{23}^2 s_{33}^2 + s_{12}^2 s_{22}^2 s_{33}^2 \end{array}\right) \\
&-20\left(\begin{array}{c} s_{11}^2 s_{13}^2 s_{23}^2 + s_{12}^2 s_{13}^2 s_{22}^2 + s_{11}^2 s_{12}^2 s_{23}^2 \\ +s_{12}^2 s_{13}^2 s_{33}^2 + s_{13}^2 s_{22}^2 s_{23}^2 + s_{12}^2 s_{23}^2 s_{33}^2 \end{array}\right) \\
&+84 s_{12}^2 s_{13}^2 s_{23}^2 + 48 s_{11}s_{12}s_{13}s_{22}s_{23}s_{33}],
\end{aligned}
$$

Proof *Solving polynomial Eq.* $\det(\mathbf{S} - \lambda\mathscr{I}_6) = 0$, *the principal curvatures are found.*

Theorem 4.4 *Let* $\mathfrak{x}$ *be a tri-rotational hypersurface determined by Eq.* (4.11). *Then,* $\mathfrak{x}$ *contains the curvatures*

$$\mathscr{K}_0 = 1,$$

$$6\mathscr{K}_1 = s_{11} + s_{22} + s_{33} + s_{44} + s_{55} + s_{66},$$

$$
\begin{aligned}
15\mathscr{K}_2 = &-\left(s_{12}^2 + s_{13}^2 + s_{23}^2\right) + s_{11}s_{22} + (s_{11} + s_{22})\, s_{33} + (s_{11} + s_{22} + s_{33})\, s_{44} \\
&+ (s_{11} + s_{22} + s_{33} + s_{44})\, s_{55} + (s_{11} + s_{22} + s_{33} + s_{44} + s_{55})\, s_{66},
\end{aligned}
$$

$$
\begin{aligned}
20\mathscr{K}_3 = &-(s_{11} + s_{22})\left(s_{13}^2 + s_{23}^2\right) + (s_{11}s_{13} + s_{12}s_{23})\, s_{13} \\
&+ (s_{12}s_{13} + s_{22}s_{23})\, s_{23} + \left(s_{11}s_{22} - s_{12}^2\right) s_{33} \\
&+ \left(s_{11}s_{22} - \left(s_{12}^2 + s_{13}^2 + s_{23}^2\right) + (s_{11} + s_{22})\, s_{33}\right) s_{44} \\
&+ \left(\begin{array}{c} s_{11}s_{22} - \left(s_{12}^2 + s_{13}^2 + s_{23}^2\right) \\ +(s_{11} + s_{22})\, s_{33} + (s_{11} + s_{22} + s_{33})\, s_{44} \end{array}\right) s_{55} \\
&+ \left(\begin{array}{c} -\left(s_{12}^2 + s_{13}^2 + s_{23}^2\right) + s_{11}s_{22} + (s_{11} + s_{22})\, s_{33} \\ +(s_{11} + s_{22} + s_{33})\, s_{44} + (s_{11} + s_{22} + s_{33} + s_{44})\, s_{55} \end{array}\right) s_{66},
\end{aligned}
$$

$$15\mathcal{K}_4=\begin{pmatrix}(s_{11}s_{13}+s_{12}s_{23})s_{13}+(s_{12}s_{13}+s_{22}s_{23})s_{23}\\-(s_{11}+s_{22})\left(s_{13}^2+s_{23}^2\right)+\left(s_{11}s_{22}-s_{12}^2\right)s_{33}\end{pmatrix}s_{44}$$
$$+\begin{pmatrix}(s_{11}s_{13}+s_{12}s_{23})s_{13}+(s_{12}s_{13}+s_{22}s_{23})s_{23}\\-(s_{11}+s_{22})\left(s_{13}^2+s_{23}^2\right)+\left(s_{11}s_{22}-s_{12}^2\right)s_{33}\\+\left(s_{11}s_{22}-\left(s_{12}^2+s_{13}^2+s_{23}^2\right)+(s_{11}+s_{22})s_{33}\right)s_{44}\end{pmatrix}s_{55}$$
$$+\begin{pmatrix}s_{13}(s_{11}s_{13}+s_{12}s_{23})+s_{23}(s_{12}s_{13}+s_{22}s_{23})\\+s_{33}\left(s_{11}s_{22}-s_{12}^2\right)+(-s_{11}-s_{22})\left(s_{13}^2+s_{23}^2\right)\\-s_{44}\left(-s_{11}s_{22}+s_{12}^2+s_{13}^2+s_{23}^2+s_{33}(-s_{11}-s_{22})\right)\\+s_{55}\begin{pmatrix}s_{11}s_{22}+s_{44}(s_{11}+s_{22}+s_{33})\\-s_{12}^2-s_{13}^2-s_{23}^2-s_{33}(-s_{11}-s_{22})\end{pmatrix}\end{pmatrix}s_{66},$$

$$6\mathcal{K}_5=\begin{pmatrix}(s_{11}s_{13}+s_{12}s_{23})s_{13}+(s_{12}s_{13}+s_{22}s_{23})s_{23}\\-(s_{11}+s_{22})\left(s_{13}^2+s_{23}^2\right)+\left(s_{11}s_{22}-s_{12}^2\right)s_{33}\end{pmatrix}s_{44}s_{55}$$
$$+\begin{pmatrix}-\begin{pmatrix}(s_{11}s_{13}+s_{12}s_{23})s_{13}\\+(s_{12}s_{13}+s_{22}s_{23})s_{23}\\+\left(s_{11}s_{22}-s_{12}^2\right)s_{33}\\-(s_{11}+s_{22})\left(s_{13}^2+s_{23}^2\right)\end{pmatrix}s_{44}\\+\begin{pmatrix}-(s_{11}s_{13}+s_{12}s_{23})s_{13}-(s_{12}s_{13}+s_{22}s_{23})s_{23}\\-\left(s_{11}s_{22}-s_{12}^2\right)s_{33}+(s_{11}+s_{22})\left(s_{13}^2+s_{23}^2\right)\\+\left(-s_{11}s_{22}+s_{12}^2+s_{13}^2+s_{23}^2-s_{33}(s_{11}+s_{22})\right)s_{44}\end{pmatrix}s_{55}\end{pmatrix}s_{66},$$

$$\mathcal{K}_6=\begin{pmatrix}(s_{11}s_{13}+s_{12}s_{23})s_{13}+(s_{12}s_{13}+s_{22}s_{23})s_{23}\\-(s_{11}+s_{22})\left(s_{13}^2+s_{23}^2\right)+\left(s_{11}s_{22}-s_{12}^2\right)s_{33}\end{pmatrix}s_{44}s_{55}s_{66}.$$

Here, $\mathcal{K}_1, \mathcal{K}_6$ denote the mean curvature H, the Gauss–Kronecker curvature K, respectively.

Proof *By using the Cayley-Hamilton theorem, we reveal the following characteristic polynomial Eq. $P_{\mathbf{S}}(\lambda)=0$ of* **S***:*

$$\mathcal{K}_0\lambda^6-6\mathcal{K}_1\lambda^5+15\mathcal{K}_2\lambda^4-20\mathcal{K}_3\lambda^3+15\mathcal{K}_4\lambda^2-6\mathcal{K}_5\lambda+\mathcal{K}_6=0.$$

The curvatures $\mathcal{K}_i$ of $\mathfrak{x}$ are obtained by the above Eq.

Corollary 4.1 *Let $\mathfrak{x}$ be a tri-rotational hypersurface defined by Eq.* (4.11). *Then, $\mathfrak{x}$ has zero mean curvature iff the following partial differential Eq. appears*

$$s_{11}+s_{22}+s_{33}+s_{44}+s_{55}+s_{66}=0.$$

Corollary 4.2 *Let $\mathfrak{x}$ be a tri-rotational hypersurface given by Eq.* (4.11). *Then, $\mathfrak{x}$ zero Gauss-Kronecker curvature iff the following partial differential equation occurs*

$$[(s_{11}s_{13}+s_{12}s_{23})s_{13}+(s_{12}s_{13}+s_{22}s_{23})s_{23}$$
$$-(s_{11}+s_{22})\left(s_{13}^2+s_{23}^2\right)+\left(s_{11}s_{22}-s_{12}^2\right)s_{33}]s_{44}s_{55}s_{66}=0.$$

Hence, we serve the following.

Example 4.1 *Consider a tri-rotational hypersurface denoted as $\mathfrak{x}$, determined by Eq.* (4.11)*. When the generating hypersurface γ of $\mathfrak{x}$ is parametrized by the unit hypersphere with functions $f(u,v,w) = \cos u \cos v \cos w$, $g(u,v,w) = \sin u \cos v \cos w$, $h(u,v,w) = \sin v \cos w$, and $k(u,v,w) = \sin w$, we find that $\mathbf{S} = \mathscr{I}_6$, and the tri-rotational hypersurface exhibits the curvatures $\mathscr{K}_i = 1$, where $i = 0, 1, ..., 6$.*

Example 4.2 *Let us consider a tri-rotational hypersurface denoted as $\mathfrak{x}$, determined by Eq.* (4.11)*. Assuming that the generating hypersurface γ of $\mathfrak{x}$ is parametrized by the rational unit hypersphere with functions $f(u,v,w) = \frac{1-u^2}{1+u^2}\frac{1-v^2}{1+v^2}\frac{1-w^2}{1+w^2}$, $g(u,v,w) = \frac{2u}{1+u^2}\frac{1-v^2}{1+v^2}\frac{1-w^2}{1+w^2}$, $h(u,v,w) = \frac{2v}{1+v^2}\frac{1-w^2}{1+w^2}$, and $k(u,v,w) = \frac{2w}{1+w^2}$, we find that the tri-rotational hypersurface yields the same results as determined in Example 1.*

Example 4.3 *Consider a tri-rotational hypersurface denoted as $\mathfrak{x}$, defined by Eq.* (4.11)*. Assuming that the generating hypersurface γ of $\mathfrak{x}$ is parametrized by the Riemann hypersphere with functions $f(u,v,w) = \frac{2u}{u^2+v^2+w^2+1}$, $g(u,v,w) = \frac{2v}{u^2+v^2+w^2+1}$, $h(u,v,w) = \frac{2w}{u^2+v^2+w^2+1}$, and $k(u,v,w) = \frac{u^2+v^2+w^2-1}{u^2+v^2+w^2+1}$, we find that the tri-rotational hypersurface has $\mathbf{S} = -\mathscr{I}_6$ and exhibits the following curvatures: $\mathscr{K}_i = (-1)^i$, where $i = 0, 1, ..., 6$.*

Example 4.4 *Consider the hypersphere $\mathbb{S}^6(r) = \{\xi \in \mathbb{E}^7 \mid \langle \xi, \xi \rangle = r^2\}$, where $r > 0$ represents the radius. Parametrizing the hypersphere as*

$$\xi(u,v,w,s,p,q) = \begin{pmatrix} r\cos u \cos v \cos w \cos s \\ r\cos u \cos v \cos w \sin s \\ r\sin u \cos v \cos w \cos p \\ r\sin u \cos v \cos w \sin p \\ r\sin v \cos w \cos q \\ r\sin v \cos w \sin q \\ r\sin w \end{pmatrix}, \tag{4.15}$$

we obtain $\mathbf{S} = \frac{1}{r}\mathscr{I}_6$. Consequently, the hypersphere defined by Eq. (4.15) *exhibits the curvatures $\mathscr{K}i = \frac{1}{r^i}$, where $i = 0, 1, ..., 6$. Moreover, in this case, we have $H^6 = K$, indicating that the hypersurface defined by Eq.* (4.15) *is a tri-rotational umbilical hypersphere.*

4.5 TRI-ROTATIONAL HYPERSURFACE SATISFYING $\Delta\mathfrak{X} = \mathscr{A}\mathfrak{X}$ IN $\mathbb{E}^7$

In this section, our focus is on the Laplace–Beltrami operator in $\mathbb{E}^7$. We will proceed to compute it utilizing the tri-rotational hypersurface, which is defined by Eq. (4.11).

By using the inverse matrix of the first fundamental form matrix $(\mathbf{g}_{ij})_{6\times 6}$, where $\hat{\mathbf{g}} = \det(\mathbf{g}_{ij})$, we have the following.

Definition 4.2 *The operator Δ is defined as the Laplace–Beltrami operator of a smooth function $\phi = \phi(x^1, x^2, ..., x^6)\,|_{\mathbf{D}}$ of class C^6*

$$\Delta\phi = \frac{1}{\hat{\mathbf{g}}^{1/2}} \sum_{i,j=1}^{6} \frac{\partial}{\partial x^i}\left(\mathbf{g}^{1/2}\mathbf{g}^{ij}\frac{\partial \phi}{\partial x^j}\right) \tag{4.16}$$

on a subset $\mathbf{D} \subset \mathbb{R}^6$. Here, $(\mathbf{g}^{ij}) = (\mathbf{g}_{kl})^{-1}$ and $\hat{\mathbf{g}} = \det(\mathbf{g}_{ij})$.

For tri-rotational hypersurface (4.11), $\mathbf{g}_{ij} = 0$ when $i \neq j$ except for $i, j < 4$. Therefore, Δ operator of hypersurface $\mathfrak{x} = \mathfrak{x}(u, v, w, s, p, q)$ is described by

$$\Delta\mathfrak{x} = \frac{1}{\hat{\mathbf{g}}^{1/2}} \left\{ \begin{array}{c} \frac{\partial}{\partial u}\left(\hat{\mathbf{g}}^{1/2}\mathbf{g}^{11}\frac{\partial \mathfrak{x}}{\partial u}\right) + \frac{\partial}{\partial u}\left(\hat{\mathbf{g}}^{1/2}\mathbf{g}^{12}\frac{\partial \mathfrak{x}}{\partial v}\right) + \frac{\partial}{\partial u}\left(\hat{\mathbf{g}}^{1/2}\mathbf{g}^{13}\frac{\partial \mathfrak{x}}{\partial w}\right) \\ + \frac{\partial}{\partial v}\left(\hat{\mathbf{g}}^{1/2}\mathbf{g}^{21}\frac{\partial \mathfrak{x}}{\partial u}\right) + \frac{\partial}{\partial v}\left(\hat{\mathbf{g}}^{1/2}\mathbf{g}^{22}\frac{\partial \mathfrak{x}}{\partial v}\right) - \frac{\partial}{\partial v}\left(\hat{\mathbf{g}}^{1/2}\mathbf{g}^{23}\frac{\partial \mathfrak{x}}{\partial w}\right) \\ + \frac{\partial}{\partial w}\left(\mathbf{g}^{1/2}\mathbf{g}^{31}\frac{\partial \mathfrak{x}}{\partial u}\right) + \frac{\partial}{\partial w}\left(\hat{\mathbf{g}}^{1/2}\mathbf{g}^{32}\frac{\partial \mathfrak{x}}{\partial v}\right) + \frac{\partial}{\partial w}\left(\hat{\mathbf{g}}^{1/2}\mathbf{g}^{33}\frac{\partial \mathfrak{x}}{\partial w}\right) \\ + \frac{\partial}{\partial s}\left(\hat{\mathbf{g}}^{1/2}\mathbf{g}^{44}\frac{\partial \mathfrak{x}}{\partial s}\right) + \frac{\partial}{\partial p}\left(\hat{\mathbf{g}}^{1/2}\mathbf{g}^{55}\frac{\partial \mathfrak{x}}{\partial p}\right) + \frac{\partial}{\partial q}\left(\hat{\mathbf{g}}^{1/2}\mathbf{g}^{66}\frac{\partial \mathfrak{x}}{\partial q}\right) \end{array} \right\}. \tag{4.17}$$

By taking the derivatives of the functions in Eq. (4.17) with respect to u, v, w, s, p, q, respectively, we can derive the following.

Theorem 4.5 *The Laplace–Beltrami operator of the tri-rotational hypersurface, denoted by Eq.* (4.11), *can be expressed as* $\Delta\mathfrak{x} = 6\mathbb{K}_1\mathbb{G}$, *where* $\mathbb{K}_1$ *represents the mean curvature and* $\mathbb{G}$ *represents the Gauss map associated with* $\mathfrak{x}$.

Proof *By directly evaluating the expression in Eq.* (4.17), *we obtain* $\Delta\mathfrak{x}$.

Theorem 4.6 *Let* $\mathfrak{x}$ *represent the tri-rotational hypersurface defined by Eq.* (4.11). *The equation* $\Delta\mathfrak{x} = \mathscr{M}\mathfrak{x}$ *holds, where* $\mathscr{M}$ *denotes a square matrix of order* 7, *if and only if* $\mathfrak{x}$ *possesses* $\mathbb{K}_1 = 0$, *indicating that it is a minimal hypersurface.*

Proof *By calculating* $6\mathbb{K}_1\mathbb{G} = \mathscr{M}\mathfrak{x}$, *we establish the following relationship*

$$\begin{aligned}
& m_{11} f\cos s + m_{12} f\sin s + m_{13} f\cos p + m_{14} f\sin p + m_{15} f\cos q + m_{16} f\sin q + m_{17} k \\
={}& \phi fghG_1 \cos s, \\
& m_{21} f\cos s + m_{22} f\sin s + m_{23} f\cos p + m_{24} f\sin p + m_{25} f\cos q + m_{26} f\sin q + m_{27} k \\
={}& \phi fghG_1 \sin s, \\
& m_{31} f\cos s + m_{32} f\sin s + m_{33} f\cos p + m_{34} f\sin p + m_{35} f\cos q + m_{36} f\sin q + m_{37} k \\
={}& \phi fghG_2 \cos p, \\
& m_{41} f\cos s + m_{42} f\sin s + m_{43} f\cos p + m_{44} f\sin p + m_{45} f\cos q + m_{46} f\sin q + m_{47} k \\
={}& \phi fghG_2 \sin p, \\
& m_{51} f\cos s + m_{52} f\sin s + m_{53} f\cos p + m_{54} f\sin p + m_{55} f\cos q + m_{56} f\sin q + m_{57} k \\
={}& \phi fghG_3 \cos q, \\
& m_{61} f\cos s + m_{62} f\sin s + m_{63} f\cos p + m_{64} f\sin p + m_{65} f\cos q + m_{66} f\sin q + m_{67} k \\
={}& \phi fghG_3 \sin q, \\
& m_{71} f\cos s + m_{72} f\sin s + m_{73} f\cos p + m_{74} f\sin p + m_{75} f\cos q + m_{76} f\sin q + m_{77} k \\
={}& \phi fghG_4,
\end{aligned}$$

where $\mathscr{M} = (m_{ij})$ *is the* 7×7 *matrix,* $\phi = 6\mathbb{K}_1\hat{\mathbf{g}}^{-1/2}$. *Derivativing above ODEs twice w.r.t. s, we obtain*

$$m_{i7} = 0, \; \phi = 0, \; i = 1, 2, ..., 7.$$

Then, we get

$$m_{i1} f\cos s + m_{i2} f\sin s = 0,$$

where $i = 1,2,...,7$. sin *and* cos *are linearly independent with respect to s, which implies that all components of the matrix* $\mathscr{M}$ *are equal to* 0. *As* $\phi = 6\mathbb{K}_1\hat{\mathbf{g}}^{-1/2}$, *we conclude that* $\mathbb{K}_1 = 0$. *This implies that* $\mathfrak{x}$ *is a minimal rotational hypersurface.*

Therefore, we give the following.

Example 4.5 *Consider the tri-rotational hypersurface denoted by* $\mathfrak{x}$, *given by Eq.* (4.11). *When the generating hypersurface* γ *of* $\mathfrak{x}$ *is parametrized by the unit hypersphere as determined in Example 4.1, the tri-rotational hypersurface* $\mathfrak{x}$ *yields* $\Delta\mathfrak{x} = \mathscr{A}\mathfrak{x}$, *where* $\mathscr{A} = -6\mathscr{I}_7$ *and* $\mathscr{I}_7$ *denotes the identity matrix.*

Example 4.6 *Let* $\mathfrak{x}$ *represent the tri-rotational hypersurface defined by Eq.* (4.11). *Assuming that the generating hypersurface* γ *of* $\mathfrak{x}$ *is parametrized by the Riemann hypersphere as defined in Example 4.3, we find that the tri-rotational hypersurface* $\mathfrak{x}$ *yields the same results as indicated in Example 4.5.*

4.6 CONCLUSIONS

We have presented a comprehensive analysis of the rotational hypersurface denoted by $\mathfrak{x}$, which is constructed through triple rotation in the seven-dimensional Euclidean space $\mathbb{E}^7$.

We have provided detailed insights into various aspects of this hypersurface, including the fundamental forms, Gauss map, and shape operator. By leveraging the Cayley–Hamilton theorem, we have computed the curvatures and principal curvatures of $\mathfrak{x}$. Additionally, we have established equations that describe the relationships between the mean curvature and Gauss–Kronecker curvature of $\mathfrak{x}$. Furthermore, we have unveiled the Laplace–Beltrami operator relation of $\mathfrak{x}$.

Overall, our research provides valuable insights into the geometric properties and mathematical characterization of this particular rotational hypersurface in $\mathbb{E}^7$.

REFERENCES

1. Alias, L. J. & Gürbüz, N. (2006). An extension of Takashi theorem for the linearized operators of the highest order mean curvatures, Geom. Dedicata **121**, 113–127.
2. Aminov, Y. (2001). The geometry of submanifolds. Gordon and Breach Sci. Pub., Amsterdam.
3. Arslan, K., Deszcz, R., & Yaprak, Ş. (1997). On Weyl pseudosymmetric hypersurfaces. Colloq. Math. **72**(2), 353–361.
4. Chen, B. Y. (1983). On submanifolds of finite type. Soochow J. Math. **9**, 65–81.
5. Chen, B. Y. (1984). Total mean curvature and submanifolds of finite type. World Scientific, Singapore.
6. Chen, B. Y. (1985). Finite type submanifolds and generalizations. University of Rome.
7. Chen, B. Y. (1985). Finite type submanifolds in pseudo-Euclidean spaces and applications. Kodai Math. J. **8**(3), 358–374.
8. Chen, B. Y., Güler, E., Yaylı, Y., Hacısalihoğlu, H. H. (2023). *Differential geometry of 1-type submanifolds and submanifolds with 1-type Gauss map*, Int. Elec. J. Geom., **16**(1), 4–49.
9. Chen, B. Y. & Piccinni, P. (1987). Submanifolds with finite type Gauss map. Bull. Austral. Math. Soc. **35**, 161–186.
10. Cheng, S. Y. & Yau, S. T. (1977). Hypersurfaces with constant scalar curvature. Math. Ann. **225**, 195–204.
11. Do Carmo, M. & Dajczer, M. (1983). Rotation Hypersurfaces in Spaces of Constant Curvature. Trans. Amer. Math. Soc. **277**, 685–709.
12. Dursun, U. (2007). Hypersurfaces with pointwise 1-type Gauss map. Taiwanese J. Math. **11**(5), 1407–1416.
13. Ferrandez, A., Garay, O. J., & Lucas, P. (1990). On a certain class of conformally at Euclidean hypersurfaces. In Global Analysis and Global Differential Geometry; Springer: Berlin, Germany 48–54.

14. Garay, O. J. (1990). An extension of Takahashi's theorem. Geom. Dedicata **34**, 105–112.
15. Güler, E., Magid, M., & Yaylı, Y. (2016). *Laplace–Beltrami operator of a helicoidal hypersurface in four-space*, J. Geom. Symmetry Phys. **41**, 77–95.
16. Güler, E., Hacısalihoğlu, H. H., & Kim, Y. H. (2018). *The Gauss map and the third Laplace–Beltrami operator of the rotational hypersurface in 4-space*, Symmetry **10**, 1–12.
17. Kühnel, W. (2015). Differential geometry. Curves-surfaces-manifolds. Third ed. Translated from the 2013 German ed. AMS, Providence, RI.
18. Levi-Civita, T. (1937). Famiglie di superficie isoparametriche nellordinario spacio euclideo. Rend. Acad. Lincei **26**, 355–362.
19. Li, Y., & Güler, E. (2023). Hypersurfaces of revolution family supplying $\Delta \mathfrak{r} = A\mathfrak{r}$ in pseudo-Euclidean space $\mathbb{E}^7_3$. AIMS Mathematics **8**(10), 24957–24970.
20. Takahashi, T. (1966). Minimal immersions of Riemannian manifolds. J. Math. Soc. Japan **18**, 380–385.

5 On Recent Developments in the Non-Inner Automorphism Conjecture

Sandeep Singh
Department of Mathematics, Akal University, Talwandi Sabo, India

Rohit Garg
Department of Mathematics, Government Ripudaman College, Nabha, India
School of Mathematical Sciences, National Institute of Science Education and Research, Bhubaneswar, India

CONTENTS

In this chapter, we will present the recent development on the non-inner automorphism conjecture.

5.1 INTRODUCTION

We denote by $\Phi(G)$ and $d(G)$, respectively, the Frattini subgroup and the rank of the group G. By $Z_i(G)$, where $i \geq 2$, we denote the ith term of the upper central series of G, where $Z_1(G) = Z(G)$, the center of G. By $\gamma_i(G)$, where $i \geq 2$, we denote the ith term of the lower central series of G, where $\gamma_2(G) = G'$, the commutator subgroup of G. For a nilpotent group G, the smallest positive integer c such that $Z_c(G) = G$ (or $\gamma_{c+1}(G) = 1$) is called the **nilpotency class** (or simply **class**) of G. For a finite p-group G, let $\Omega_i(G) = \langle g \in G \mid g^{p^i} = 1 \rangle$, where i is a positive integer. The following is a long-standing conjecture:

Non-inner Automorphism Conjecture: *Every finite non-abelian p-group admits an automorphism of order p which is not an inner.*

This conjecture is listed as an unsolved problem in the renowned book "Unsolved Problems in Group Theory: The Kourovka Notebook" (see [22, Problem 4.13]). The conjecture stems from Gaschutz's result that finite p-groups above order p possess non-inner automorphisms of p-power order [14]. The conjecture was first attacked by Lieback [23] in 1965. He proved the following result:

Theorem 5.1 *A finite p-group G of class 2 has a non-inner automorphism of order p if p is an odd prime, and order 2 or 4 if p = 2. Moreover, such an automorphism can be chosen leaving $\Phi(G)$ element-wise fixed.*

DOI: 10.1201/9781032634142-5

In 2002, Deaconescu and Silberberg [10] proved that if G is a finite p-group such that $C_G(Z(\Phi(G))) \neq \Phi(G)$, then G has a non-inner automorphism of order p leaving $\Phi(G)$ element-wise fixed. In 2007, Abdollahi [1] proved the conjecture for finite p-groups of class 2 by using this result and pointed out the following revised version of Liebeck's result for class 2 when $p = 2$.

Theorem 5.2 *Every finite non-abelian p-group G of class 2 has a non-inner automorphism of order p leaving either $\Phi(G)$ or $\Omega_1(Z(G))$ element-wise fixed.*

In 2013, Abdollahi et al. [6] proved the conjecture for all finite p-groups of class 3. A finite p-group G is called **regular** if for any two elements $x, y \in G$, there exists an element c in the commutator subgroup $[\langle x,y\rangle, \langle x,y\rangle]$ of $\langle x,y\rangle$ such that $(xy)^p = x^p y^p c^p$. In 1980, Schmid [28] proved the conjecture for all regular p-groups. It is well-known that for an odd prime p, every finite p-group G with cyclic G' is regular. It follows that if G is a finite p-group ($p > 2$) with cyclic G', then G satisfies the conjecture. For finite 2-groups, Jamali and Viseh [21] proved the following result:

Theorem 5.3 *Let G be a finite non-abelian 2-group such that G' is cyclic. Then G has a non-inner automorphism of order 2 fixing either $\Phi(G)$ or $Z(G)$ element-wise.*

Let G and H be two groups. By $Hom(G,H)$, we denote the set of all group homomorphisms from G to H. Observe that if H is abelian, then $Hom(G,H)$ forms an abelian group under the binary operation defined by $\alpha\beta(g) = \alpha(g)\beta(g)$ for all $\alpha, \beta \in Hom(G,H)$ and for all $g \in G$. For normal subgroups X and Y of G, let $Aut^X(G)$ and $Aut_Y(G)$, respectively, denote the subgroups of $Aut(G)$ centralizing G/X and Y. We denote the intersection $Aut^X(G) \cap Aut_Y(G)$ by $Aut_Y^X(G)$. The following lemma is a little modification of arguments of [8, Lemma 3].

Lemma 5.1 *Let G be any group and X be a central subgroup of G contained in a normal subgroup Y of G. Then $Aut_Y^X(G) \simeq Hom(G/Y, X)$.*

An automorphism α of G is called **central** if it centralizes $G/Z(G)$, or equivalently, $g^{-1}\alpha(g) \in Z(G)$ for all $g \in G$. By Lemma 12.1, we have the group $Aut_{\Phi(G)}^{Z(G)}(G)$ of all central automorphisms of G which fixes $\Phi(G)$ element-wise is isomorphic to $Hom(G/\Phi(G), Z(G))$ in case $Z(G) \leq \Phi(G)$.

Lemma 5.2 ([29, Proposition 2.2]) *Let G be a finite non-abelian p-group such that $Aut_{\Phi(G)}^{Z(G)}(G)$ is contained in $Inn(G)$. Then*

(1) $Z(G) \leq \Phi(G)$.
(2) $\frac{Z_2(G) \cap C_G(\Phi(G))}{Z(G)} \simeq Hom(G/\Phi(G), G' \cap Z(G))$.
(3) $d(G' \cap Z(G)) = d(Z(G))$.

Proof *(1) Assume that $Z(G)$ is not contained in $\Phi(G)$. Let $g_0 \in Z(G) - M$ for some maximal subgroup M of G and let $z_0 \in Z(G) \cap \Phi(G)$ be of order p. Then, the map $\alpha : G \to G$ defined as $\alpha(mg_0^i) = mg_0^i z_0^i$ for all $m \in M$ is easily seen to be a central automorphism of order p leaving $\Phi(G)$ element-wise fixed. By assumption, we have $\alpha = \theta_a$, the inner automorphism induced by some $a \in G$. Then $g_0 z_0 = \alpha(g_0) = \theta_a(g_0) = g_0$. It follows that $z_0 = 1$, a contradiction. Hence $Z(G) \leq \Phi(G)$.*

(2) Let $x \in Z_2(G) \cap C_G(\Phi(G))$ be any element. Then, it is easy to check that θ_x, the inner automorphism induced by x, belongs to $Aut_{\Phi(G)}^{Z(G)}(G)$. Also, the map $\alpha : Z_2(G) \cap C_G(\Phi(G)) \to Aut_{\Phi(G)}^{Z(G)}(G)$ defined as $\alpha(x) = \theta_{x^{-1}}$ is an epimorphism with kernel $Z(G)$. Thus

$$Aut_{\Phi(G)}^{Z(G)}(G) \simeq \frac{Z_2(G) \cap C_G(\Phi(G))}{Z(G)}.$$

Define the map $\sigma : Aut_{\Phi(G)}^{Z(G)}(G) \to Hom(G/\Phi(G), G' \cap Z(G))$ as $\sigma(\tau) = \sigma_\tau$ for all $\tau \in Aut_{\Phi(G)}^{Z(G)}(G)$, where $\sigma_\tau(x\Phi(G)) = x^{-1}\tau(x)$. Then, it is easy to verify that σ is a monomorphism. Now, we prove that σ is surjective. Let $f \in Hom(G/\Phi(G), G' \cap Z(G))$ be any element. Then, the map β defined as $\beta(x) = xf(x\Phi(G))$ for all $x \in G$ is an automorphism of G. Since $x^{-1}\beta(x) = f(x\Phi(G)) \in Z(G)$ for all $x \in G$ and $\beta(x) = x$ for all $x \in \Phi(G)$, we have $\beta \in Aut_{\Phi(G)}^{Z(G)}(G)$. Hence

$$Hom(G/\Phi(G), G' \cap Z(G)) \simeq Aut_{\Phi(G)}^{Z(G)}(G) \simeq \frac{Z_2(G) \cap C_G(\Phi(G))}{Z(G)}.$$

(3) Since $Z(G) \leq \Phi(G)$, we have $Aut_{\Phi(G)}^{Z(G)}(G) \simeq Hom(G/\Phi(G), Z(G))$. Therefore, it follows from part (2) that

$$Hom(G/\Phi(G), G' \cap Z(G)) \simeq Hom(G/\Phi(G), Z(G)),$$

and hence $d(G' \cap Z(G)) = d(Z(G))$, because $G/\Phi(G)$ is elementary abelian.

The main result in [29] is the following Theorem:

Theorem 5.4 *Let G be a finite non-abelian p-group. If G satisfies one of the following conditions*

(1) $d(G' \cap Z(G)) \neq d(Z(G))$, or
(2) $\frac{Z_2(G)}{Z(G)}$ is cyclic, or
(3) $C_G(Z(\Phi(G))) = \Phi(G)$ and $\frac{Z_2(G) \cap Z(\Phi(G))}{Z(G)}$ is not elementary abelian of rank rs, where $r = d(G)$ and $s = d(Z(G))$,

then G has a non-inner central automorphism of order p which fixes $\Phi(G)$ element-wise.

Proof *First, assume that $Z(G) \nleq \Phi(G)$. Then, by the proof of part (1) of Lemma 12.2, we have G has a non-inner central automorphism of order p leaving $\Phi(G)$ element-wise fixed. Now, assume that $Z(G) \leq \Phi(G)$. Thus $Aut_{\Phi(G)}^{Z(G)}(G) \simeq Hom(G/\Phi(G), Z(G))$. We prove that $Aut_{\Phi(G)}^{Z(G)}(G) \nleq Inn(G)$ holds under any of the three specified hypothesis conditions. Firstly, if $d(G' \cap Z(G)) \neq d(Z(G))$, then $Aut_{\Phi(G)}^{Z(G)}(G) \nleq Inn(G)$ by Lemma 12.2. Next, if $Z_2(G)/Z(G)$ is cyclic, then $\frac{Z_2(G) \cap C_G(\Phi(G))}{Z(G)}$ is also cyclic. Therefore $\frac{Z_2(G) \cap C_G(\Phi(G))}{Z(G)}$ is not isomorphic to $Hom(G/\Phi(G), G' \cap Z(G))$ because $Hom(G/\Phi(G), G' \cap Z(G))$ is elementary abelian of rank greater than or equal to 2. Thus $Aut_{\Phi(G)}^{Z(G)}(G) \nleq Inn(G)$ by Lemma 12.2.*

Finally, assume that $C_G(Z(\Phi(G))) = \Phi(G)$ and $\frac{Z_2(G) \cap Z(\Phi(G))}{Z(G)}$ is not elementary abelian of rank rs, where $r = d(G)$ and $s = d(Z(G))$. It follows from $C_G(Z(\Phi(G))) = \Phi(G)$ that $C_G(\Phi(G)) = Z(\Phi(G))$. Therefore $\frac{Z_2(G) \cap C_G(\Phi(G))}{Z(G)} = \frac{Z_2(G) \cap Z(\Phi(G))}{Z(G)}$ is not elementary abelian of rank rs. It follows that $\frac{Z_2(G) \cap C_G(\Phi(G))}{Z(G)}$ is not isomorphic to $Hom(G/\Phi(G), G' \cap Z(G))$. Therefore $Aut_{\Phi(G)}^{Z(G)}(G) \nleq Inn(G)$ by Lemma 12.2. Thus, for any of the three specified hypothesis conditions, we have $Aut_{\Phi(G)}^{Z(G)}(G) \nleq$

$Inn(G)$, and hence G has a non-inner central automorphism of order p leaving $\Phi(G)$ element-wise fixed.

A finite p-group G is called **powerful** if $G' \leq G^p$ (for $p > 2$) and $G' \leq G^4$ (for $p = 2$). In 2010, Abdollahi [2] proved the conjecture for all finite non-abelian p-groups G with $G/Z(G)$ is powerful. For a subgroup H of G, let x^H denote the subset $\{h^{-1}xh \mid h \in H\}$ of G. Let G be a finite p-group and let N be a non-trivial proper normal subgroup of G. The pair (G,N) is called a **Camina pair** if $xN \subseteq x^G$ for all $x \in G - N$. In 2012, Ghoraishi [16] proved the following result:

Theorem 5.5 *Let G be a finite p-group, where p is an odd prime, such that $(G,Z(G))$ is a Camina pair. Then G has a non-inner automorphism of order p leaving $\Phi(G)$ element-wise fixed.*

In the same year, Shabani-Attar [30] proved the conjecture for all finite non-abelian p-groups of order p^m and exponent p^{m-2}. A finite p-group G of order p^n and of nilpotency class $n-1$ is said to be of **maximal class**. In the aforementioned paper, he also proved that if G is a finite p-group of maximal class, then G has at least $p(p-1)$ non-inner automorphisms of order p that fixing $\Phi(G)$ element-wise. Let $Z_2^*(G)/Z(G) = \Omega_1(Z_2(G)/Z(G))$. In [10], the authors reduced the verification of the conjecture to the degenerate case in which $C_G(Z(\Phi(G))) = \Phi(G)$. In 2013, Ghoraishi [17] improved this reduction and proved the following result:

Theorem 5.6 *If G fails to fulfil the condition $Z_2^*(G) \leq C_G(Z_2^*(G)) = \Phi(G)$, then G has a non-inner automorphism of order p leaving $\Phi(G)$ element-wise fixed.*

In the same year, Abdollahi and Ghoraishi [3] proved the following result:

Theorem 5.7 *Let G be a finite non-abelian p-group satisfying one of the following conditions:*

(1) G is regular;
(2) G is nilpotent of class 2;
(3) the commutator subgroup of G is cyclic;
(4) $G/Z(G)$ is powerful.

Then G has a non-inner automorphism of order p leaving $Z(G)$ element-wise fixed.

A finite p-group G of order p^n and of nilpotency class $n-c$ is said to be of **co-class** c. In 2014, Fouladi and Orfi [11] proved the conjecture for all finite p-groups of co-class 2 with $p > 2$. In the same year, Abdollahi et al. [7] proved the following result:

Theorem 5.8 *Every finite p-group G of co-class 2 has a non-inner automorphism of order p leaving $Z(G)$ element-wise fixed.*

In 2016, Ruscitti et al. [27] proved the conjecture for all non-abelian finite p-groups of co-class 3, where p is a prime integer such that $p \neq 3$. Recently, Patali [25] proved the conjecture for all non-abelian finite p-groups of co-class 4 and 5, where p is a prime integer such that $p \neq 5$. A finite p-group G is called **semi-abelian** if the following property holds in G:

$$(xy^{-1})^{p^n} = 1 \Longleftrightarrow x^{p^n} = y^{p^n} \text{ for any postive integer } n.$$

In 2014, Benmoussa and Guerboussa [9] proved the conjecture for all finite semi-abelian p-groups. In the same year, Abdollahi and Ghoraishi [4] proved the following result:

Theorem 5.9 *Let G be a finite 2-group. If G is of co-class 2 or if $(G,Z(G))$ is a Camina pair, then G admits a non-inner automorphism of order 2 or 4 leaving $\Phi(G)$ element-wise fixed.*

Recall that in a group G an **anti-chain** is a set of mutually incomparable elements in the lattice of its normal subgroups. A finite p-group G is called **thin** if every anti-chain in G contains at most $p+1$ elements. In 2016, Ruscitti and Legarreta [26] proved the conjecture for all finite thin p-groups. In the same year, Abdollahi and Ghoraishi [5] proved the conjecture for all 2-generator finite p-groups with abelian Frattini subgroup. In 2017, Ghoraishi [18] proved the following result:

Theorem 5.10 *Let G be a finite non-abelian p-group. If $\{a \in Z(\Phi(G)) \mid a^{2p} \in Z(G)\} \leq Z_2(G)$, then G has a non-inner automorphism σ inducing the identity on $\Phi(G)$. Moreover, σ can be chosen to have order p if $p > 2$, and 2 or 4 if $p = 2$.*

In the same year, Orfi and Fouladi [24] proved the conjecture for all finite p-groups G with $p > 2$ and $(|Z(G)|, |Z_3(G)|) \in \{(p, p^4), (p^2, p^5)\}$. Let G be a finite p-group and let $c(G)$ be its nilpotency class. We say that G is **normally constrained** if for every $i = 1, 2, ..., c(G)$, the following equivalent conditions hold true:

(i) $\gamma_i(G)$ is unique of its order, for all $i = 1, 2, ..., c(G)$,
(ii) if $N \lhd G$, then $N \leq \gamma_i(G)$ or $N \geq \gamma_i(G)$, for all $i = 1, 2, ..., c(G)$,
(iii) if $x \in G - \gamma_i(G)$, then $\gamma_i(G) \leq \langle x \rangle^G$, where H^G denotes $\bigcap_{H \leq N \unlhd G} N$.

In 2019, Gavioli et al. [15] proved the conjecture for all normally constrained p-groups when p is an odd prime. In the same year, Ghoraishi [19] proved the following result:

Theorem 5.11 *Every finite non-abelian p-group G in which $|\Phi(G)| \leq p^5$ admits a non-inner automorphism of order p leaving $Z(G)$ element-wise fixed.*

In 2020, Fouladi and Orfi [12] proved the conjecture for all finite non-abelian d-generator p-groups G with $p > 2$ and $|Z_3(G)/Z(G)| \leq p^{d+1}$. In 2021, Ghoraishi [20] proved the conjecture for all finite non-abelian p-groups G whose co-class is less than or equal to $d(G)$. In the same year, Garg and Singh [13] proved the following result:

Theorem 5.12 *Let G be a finite non-abelian p-group such that the subgroup $\{z \in Z_2(G) \mid z^p \in Z(G)\}$ of $Z_2(G)$ is non-abelian and $Z(G)$ is not cyclic. Then, there exists a non-inner automorphism of order p which fixes $\Phi(G)$ element-wise.*

Proof *Let $W(G) = \{z \in Z_2(G) \mid z^p \in Z(G)\}$ be non-abelian. Assume that every automorphism of order p which fixes $\Phi(G)$ element-wise is inner. We first prove that $C_G(W(G)) = \Phi(G)$. Let M be a maximal subgroup of G. Let $z_0 \in Z(G) \cap M$ be of order p and let $g_0 \in G - M$. Then, the map $\alpha : G \to G$ defined as $\alpha(mg_0^i) = mg_0^i z_0^i$ for all $m \in M$ is easily seen to be an automorphism of order p leaving $\Phi(G)$ element-wise fixed. By assumption, we have $\alpha = \theta_{a_M}$, the inner automorphism induced by some $a_M \in G$. It is easy to verify that $a_M \in W(G)$ and $M = C_G(a_M)$. Since $[W(G), \Phi(G)] = 1$, we have $\Phi(G) \leq C_G(W(G))$. It thus follows that*

$$\Phi(G) \leq C_G(W(G)) \leq \bigcap_M C_G(a_M) = \bigcap_M M = \Phi(G),$$

and hence $C_G(W(G)) = \Phi(G)$.

It follows from [10, Remark 1] that $Z(G) \leq \Phi(G)$. For $x \in G$ and $H \leq G$, let $\overline{x}$ and $\overline{H}$ denote the coset $x\Phi(G)$ and the quotient group $H\Phi(G)/\Phi(G)$, respectively. For each $w \in W(G)$, define the map $f_w : \overline{W(G)} \to \Omega_1(Z(G))$ as $f_w(\overline{v}) = [v,w]$ for all $v \in W(G)$. Since

$$\begin{aligned} f_w(\overline{u}\,\overline{v}) &= f_w(\overline{uv}) = [uv,w] \\ &= [u,w][u,w,v][v,w] = [u,w][v,w] \\ &= f_w(\overline{u})f_w(\overline{v}) \end{aligned}$$

for all $\overline{u},\overline{u} \in \overline{W}$, we have f_w is a group homomorphism. Define the map $f : W(G) \to Hom(\overline{W(G)},\Omega_1(Z(G)))$ as $f(w) = f_w$ for all $w \in W(G)$. Let $u,v,w \in W$ be any elements. Then $f_{uv}(\overline{w}) = [w,uv] = [w,v][w,u][w,u,v] = [w,u][w,v] = f_u(\overline{w})f_v(\overline{w})$. Thus f is a group homomorphism with $ker(f) = W(G) \cap C_G(W(G)) = W(G) \cap \Phi(G)$. We prove that f is surjective. Let $\alpha \in Hom(\overline{W(G)},\Omega_1(Z(G)))$. Since $\overline{G}$ is elementary abelian, the homomorphism α can be extended to the homomorphism $\beta : \overline{G} \to \Omega_1(Z(G))$ defined by

$$\beta(\overline{w}) = \begin{cases} \alpha(\overline{w}) & \text{if} \quad \overline{w} \in \overline{W(G)} \\ 1 & \text{if} \quad \overline{w} \in \overline{G} - \overline{W(G)}. \end{cases}$$

Since $\Omega_1(Z(G)) \leq \Phi(G)$, the map $\sigma_\beta : G \to G$ defined as $\sigma_\beta(g) = g\beta(\overline{g})$ is easily seen to be an automorphism of order p leaving $\Phi(G)$ element-wise fixed. By assumption, we have $\sigma_\beta = \theta_a$ for some $a \in W(G)$. Therefore, for each $w \in W(G)$, we have $\alpha(\overline{w}) = \beta(\overline{w}) = w^{-1}\sigma_\beta(w) = w^{-1}\theta_a(w) = [w,a] = f_a(\overline{w})$, and hence f is an epimorphism. It follows that

$$\frac{W(G)}{W(G) \cap \Phi(G)} \simeq \overline{W(G)} \simeq Hom(\overline{W(G)},\Omega_1(Z(G))),$$

and therefore $d(Z(G)) = 1$, which is a contradiction to the hypothesis. Hence, there exists a non-inner automorphism of order p leaving $\Phi(G)$ element-wise fixed.

REFERENCES

1. A. Abdollahi, *Finite p-groups of class 2 have noninner automorphisms of order p*, J. Algebra, **312**(2) (2007) 876-879.
2. A. Abdollahi, *Powerful p-groups have non-inner automorphisms of order p and some cohomology*, J. Algebra, **323**(3) (2010) 779-789.
3. A. Abdollahi and S.M. Ghoraishi, *Noninner automorphisms of finite p-groups leaving the center elementwise fixed*, Int. J. Group Theory, **2**(4) (2013) 17-20.
4. A. Abdollahi and S.M. Ghoraishi, *On non-inner 2-automorphisms of finite 2-groups*, Bull. Aust. Math. Soc., **90**(2) (2014) 227-231.
5. A. Abdollahi and S.M. Ghoraishi, *On noninner automorphisms of 2-generator finite p-groups*, Comm. Algebra, **45**(8) (2017) 3636-3642.
6. A. Abdollahi, M. Ghoraishi and B. Wilkens, *Finite p-groups of class 3 have noninner automorphisms of order p*, Beitr. Algebra Geom., **54**(1) (2013) 363-381.
7. A. Abdollahi, S.M. Ghoraishi, Y. Guerboussa, M. Reguiat and B. Wilkens, *Noninner automorphisms of order p for finite p-groups of coclass* 2, J. Group Theory, **17**(2) (2014) 267-272.
8. J.L. Alperin, *Groups with finitely many automorphisms*, Pacific J. Math., **12**(1) (1962), 1-5.
9. M.T. Benmoussa and Y. Guerboussa, *Some properties of semi-abelian p-groups*, Bull. Aust. Math. Soc., **91**(1) (2015) 86-91.
10. M. Deaconescu and G. Silberberg, *Noninner automorphisms of order p of finite p-groups*, J. Algebra, **250**(1) (2002) 283-287.
11. S. Fouladi and R. Orfi, *Noninner automorphisms of order p in finite p-groups of coclass 2, when p > 2*, Bull. Aust. Math. Soc., **90**(2) (2014) 232-236.
12. S. Fouladi and R. Orfi, *A note on the existence of noninner automorphisms of order p in some finite p-groups*, J. Algebra Appl., **20**(4) (2021) 2150067.

13. R. Garg and M. Singh, *Finite p-groups with non-cyclic center have a non-inner automorphism of order p*, Arch. Math. (Basel), **117**(2) (2021) 129-132.
14. W. Gaschutz, *Nichtabelsche p-Gruppen besitzen aussere p-Automorphismen*, J. Algebra, **4**(1) (1966) 1-2.
15. N. Gavioli, L. Legarreta and M. Ruscitti, *A note on non-inner automorphisms in finite normally constrained p-groups*, Publ. Math. Debr., **94**(1-2) (2019) 187-195.
16. S.M. Ghoraishi, *A note on automorphisms of finite p-groups*, Bull. Aust. Math. Soc., **87**(1) (2013) 24-26.
17. S.M. Ghoraishi, *On noninner automorphisms of finite nonabelian p-groups*, Bull. Aust. Math. Soc., **89**(2) (2014) 202-209.
18. S.M. Ghoraishi, *On noninner automorphisms of finite p-groups that fix the Frattini subgroup elementwise*, J. Algebra Appl., **17**(1) (2018) 1850137.
19. S.M. Ghoraishi, *On noninner automorphisms of finite p-groups that fix the center elementwise*, Int. J. Group Theory, **8**(1) (2019) 1-9.
20. S.M. Ghoraishi, *Noninner automorphisms of order p for finite p-groups of restricted coclass*, Arch. Math. (Basel), **117**(4) (2021) 361-368.
21. A.R. Jamali and M. Viseh, *On the existence of noninner automorphisms of order two in finite 2-groups*, Bull. Aust. Math. Soc., **87**(2) (2013) 278-287.
22. E.I. Khukhro and V.D. Mazurov, *Unsolved Problems in Group Theory: The Kourovka Notebook*, No. 20, Russian Academy of Sciences, Siberian Branch, Sobolev Institute of Mathematics, Novosibirsk (2022).
23. H. Liebeck, *Outer automorphisms in nilpotent p-groups of class 2*, J. London Math. Soc., **40**(1) (1965) 268-275.
24. R. Orfi and S. Fouladi, *Noninner automorphisms of order p in some finite p-groups*, Bull. Iran. Math. Soc., **44**(5) (2018) 1295-1301.
25. K. Patali, *Non-inner automorphisms of order p in finite p-groups of coclass 4 and 5*, Int. J. Group Theory, **13**(1) (2024) 97-114.
26. M. Ruscitti and L. Legarreta, *The existence of non-inner automorphisms of order p in finite thin p-groups*, (2016). https://doi.org/10.48550/arXiv.1604.07267.
27. M. Ruscitti, L. Legarreta and M.K. Yadav, *Non-inner automorphisms of order p in finite p-groups of coclass 3*, Monatsh. Math., **183**(4) (2017) 679-697.
28. P. Schmid, *A cohomological property of regular p-groups*, Math. Z., **175**(1) (1980) 1-3.
29. M. Shabani-Attar, *On a conjecture about automorphisms of finite p-groups*, Arch. Math. (Basel), **93**(5) (2009) 399-403.
30. M. Shabani-Attar, *Existence of noninner automorphisms of order p in some finite p-groups*, Bull. Aust. Math. Soc., **87**(2) (2013) 272-277.

6 On Some Properties of Modular and Homomorphic Products of Graphs

Bobin George
Department of Mathematics, Pavanatma College, Murickassery, Kerala, India

Jinta Jose
Department of Science and Humanities, Viswajyothi College of Engineering and Technology, Vazhakulam, Kerala, India

Rajesh K. Thumbakara
Department of Mathematics, Mar Athanasius College (Autonomous), Kothamangalam, Kerala, India

CONTENTS

The generation of a structure's configuration is the initial stage in a study of it. There are various methods available for this. One example of such a tool is the use of graph products. Modular and Homomorphic products are two different graph products. In this paper, we investigate some properties of these product operations like the overall vertex and edge counts and the sum of vertex degrees.

6.1 INTRODUCTION

Graphs are incredibly helpful data structures for describing relationships in the actual world. Graphs are applied to a wide range of practical issues because of their capacity to provide an abstraction to situations in the real world. Graph product is a technique which is used to combine two graphs, $G_a = (V_a, E_a)$ and $G_b = (V_b, E_b)$, to produce a graph G with the properties mentioned below: The Cartesian product $V_a \times V_b$ is the vertex set of G. Two vertices (v_1, v_2) and (v_1', v_2') of G are linked by an edge If and only if certain requirements about v_1, v_1' in G_a and/or v_2, v_2' in G_b are met. Modular product and homomorphic product are two types of graph products. In this paper, we investigate some properties of these product operations like the overall count of vertices and edges and the sum of vertex degrees.

DOI: 10.1201/9781032634142-6

6.2 PRELIMINARIES

Definition 6.1 *Consider two graphs, $G_a = (V_a, E_a)$ and $G_b = (V_b, E_b)$. Then the modular product of G_a and G_b denoted by $G_a \bigcirc G_b$ is a graph having set of vertices $V(G_a \bigcirc G_b) = V_a \times V_b$ and set of edges $E(G_a \bigcirc G_b)$, where two vertices (v_1, v_1') and (v_2, v_2') in $G_a \bigcirc G_b$ are adjacent if and only if*

1. v_1 and v_2 are adjacent in G_a and v_1' and v_2' are adjacent in G_b or
2. v_1 and v_2 are not adjacent in G_a and v_1' and v_2' are not adjacent in G_b.

Definition 6.2 *Consider two graphs, $G_a = (V_a, E_a)$ and $G_b = (V_b, E_b)$. Then the homomorphic product of G_a and G_b, denoted by $G_a \ltimes G_b$ is a graph having set of vertices $V(G_a \ltimes G_b) = V_a \times V_b$ and set of edges $E(G_a \ltimes G_b)$, where two vertices (v_1, v_1') and (v_2, v_2') in $G_a \ltimes G_b$ are adjacent if and only if*

1. $v_1 = v_2$ or
2. v_1 and v_2 are adjacent in G_a and v_1' and v_2' are not adjacent in G_b.

6.3 SOME PROPERTIES OF MODULAR PRODUCTS

Theorem 6.1 *Consider two graphs, $G_a = (V_a, E_a)$ and $G_b = (V_b, E_b)$. Then the modular product of G_a and G_b denoted by $G_a \bigcirc G_b$ contains $|V_a||V_b|$ vertices and $2\left(\frac{|V_a|(|V_a|-1)}{2} - |E_a|\right)\left(\frac{|V_b|(|V_b|-1)}{2} - |E_b|\right) + 2|E_a||E_b|$ edges.*

Proof *The vertex set of $G_a \bigcirc G_b$ is $V_a \times V_b$ which contains $|V_a||V_b|$ elements. Therefore, the overall vertex count in $G_a \bigcirc G_b$ is $|V_a||V_b|$. Also, two vertices (v_j, v_k) and (v_l, v_m) in $G_a \bigcirc G_b$ are adjacent if and only if*

1. *v_j and v_l are adjacent in G_a and v_k and v_m are adjacent in G_b or*
2. *v_j and v_l are not adjacent in G_a and v_k and v_m are not adjacent in G_b.*

Every edge in $G1 \bigcirc G2$ was created by only one of these two conditions, and neither of them can be true simultaneously. Therefore, we add the number of edges produced by each condition to obtain the overall edge count in $G_a \bigcirc G_b$. Consider the first situation for adjacency, i.e.,v_j and v_l are adjacent in G_a and v_k and v_m are adjacent in G_b. We know, $|E_a|$ and $|E_b|$ edges occur in G_a and G_b, respectively. Therefore, there are $|E_a||E_b|$ different ways to choose a couple of edges e_a and e_b so that one comes from G_a while the other comes from G_b. Assume that the edge e_a joins v_j and v_l in G_a and the edge e_b joins v_k and v_m in G_b. Then, this couple of edges yields two edges, one of which joins the vertices (v_j, v_k) and (v_l, v_m) and the other of which joins the vertices (v_j, v_m) and (v_l, v_k). In other words, we obtain $2|E_a||E_b|$ edges that satisfy the first requirement of adjacency. Take into account the second requirement for adjacency, which states that v_j and v_l are not adjacent in G_a and v_k and v_m are not adjacent in G_b. The two vertices v_j and v_l in G_a can be chosen in such a way that they are not adjacent in the graph G_a in $\left(\frac{|V_a|(|V_a|-1)}{2} - |E_a|\right)$ different ways. Similarly, we can choose two different vertices v_k and v_m in G_b such that v_k and v_m are not adjacent in G_b in $\left(\frac{|V_b|(|V_b|-1)}{2} - |E_b|\right)$ different ways. Let v_j and v_l be two vertices in G_a such that v_j and v_l are not adjacent in G_a and let v_k and v_m be two vertices in G_b such that v_k and v_m are not adjacent in G_b. As a result, we obtain two edges in $G_a \bigcirc G_b$, one of which connects the vertices (v_j, v_k) and (v_l, v_m) and the other of which connects the vertices (v_j, v_m) and (v_l, v_k). Hence totally the second condition for adjacency gives $2\left(\frac{|V_a|(|V_a|-1)}{2} - |E_a|\right)\left(\frac{|V_b|(|V_b|-1)}{2} - |E_b|\right)$ edges in $G_a \bigcirc G_b$. Therefore, the overall edge count in $G_a \bigcirc G_b$ is $2\left(\frac{|V_a|(|V_a|-1)}{2} - |E_a|\right)\left(\frac{|V_b|(|V_b|-1)}{2} - |E_b|\right) + 2|E_a||E_b|$.

Example 6.1 *Let $G_a = (V_a, E_a)$ be a graph having the set of vertices $V_a = \{v_1, v_2, v_3, v_4, v_5\}$ and the set of edges $E_a = \{v_1v_3, v_2v_3, v_2v_5, v_3v_4\}$ which is shown in the Figure 6.1.*

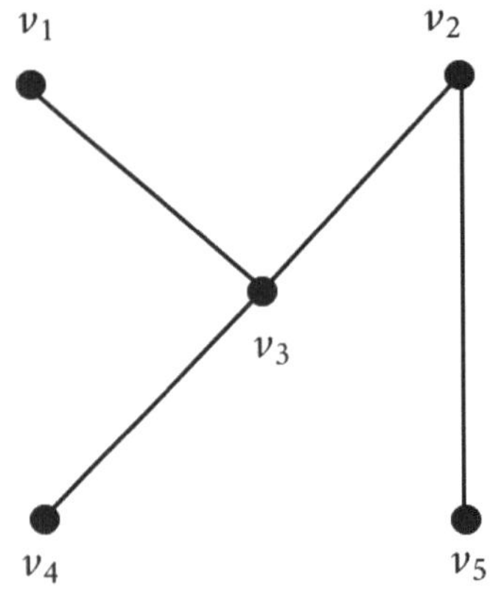

Figure 6.1 Graph $G_a = (V_a, E_a)$.

Consider another graph $G_b = (V_b, E_b)$ having vertex set $V_b = \{u_1, u_2, u_3, u_4\}$ and edge set $E_b = \{u_1u_3, u_2u_3, u_3u_4\}$ which is shown in Figure 6.2.

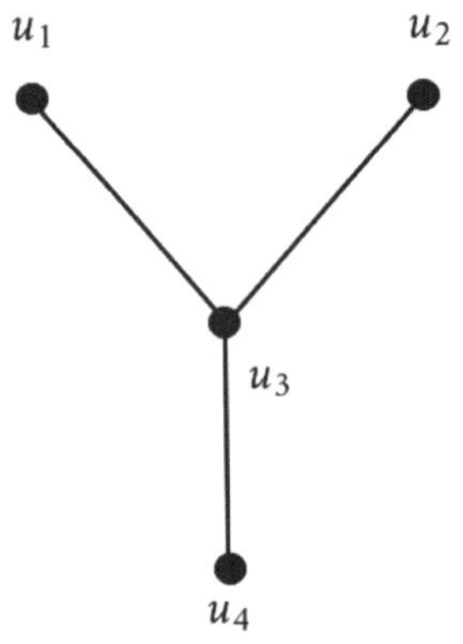

Figure 6.2 Graph $G_b = (V_b, E_b)$.

Then the modular product of these two graphs G_a and G_b is given by $G = G_a \bigcirc G_b$ and is shown in Figure 6.3.
Consider the graphs G_a and G_b and their modular product $G_a \bigcirc G_b$. Here we have, the overall count of vertices in $G_a \bigcirc G_b$ = 20 and

$$|V_a||V_b| = 5.4 = 20.$$

That is, the overall count of vertices in $G_a \bigcirc G_b = |V_a||V_b|$. Also, the overall count of edges in $G_a \bigcirc G_b = 60$ and

$$2\left(\frac{|V_a|(|V_a|-1)}{2} - |E_a|\right)\left(\frac{|V_b|(|V_b|-1)}{2} - |E_b|\right) +$$

$$2|E_a||E_b| = 2.[(5.4)/2-4][(4.3)/2-3]+2.4.3 = 60.$$

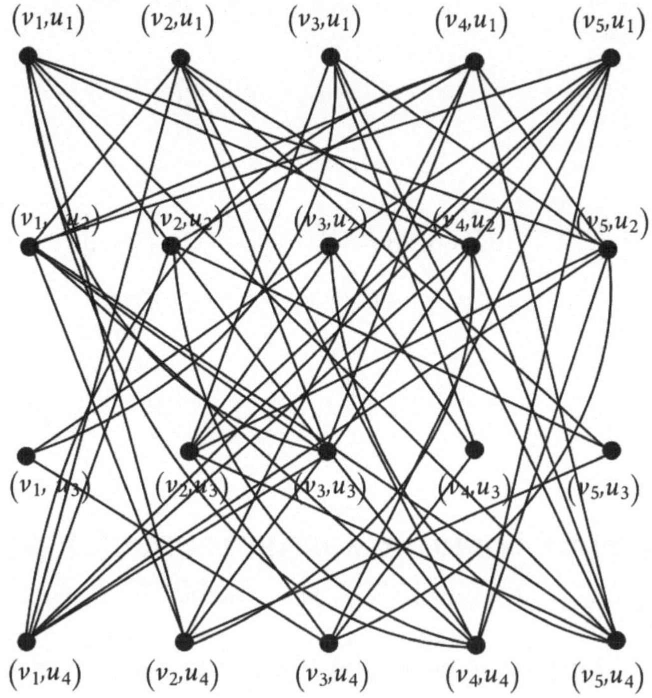

Figure 6.3 $G = G_a \bigcirc G_b$.

That is, the overall count of edges in $G_a \bigcirc G_b$=

$$2\left(\frac{|V_a|(|V_a|-1)}{2} - |E_a|\right)\left(\frac{|V_b|(|V_b|-1)}{2} - |E_b|\right) + 2|E_a||E_b|$$

.

Theorem 6.2 *Let* $G_a = (V_a, E_a)$ *and* $G_b = (V_b, E_b)$ *be two graphs. Then*

$$\sum_{(u,v)\in V_{G_a \bigcirc G_b}} deg(u,v) = 4\left(\frac{|V_a|(|V_a|-1)}{2} - |E_a|\right)\left(\frac{|V_b|(|V_b|-1)}{2} - |E_b|\right) + 4|E_a||E_b|,$$

where $deg(u,v)$ *denotes the degree of the vertex* (u,v) *in the graph* $G_a \bigcirc G_b$.

Proof *Consider the graph* $G_a \bigcirc G_b$. *By Theorem 6.1, we have the overall count edges in* $G_a \bigcirc G_b$ *is*

$$2\left(\frac{|V_a|(|V_a|-1)}{2} - |E_a|\right)\left(\frac{|V_b|(|V_b|-1)}{2} - |E_b|\right) + 2|E_a||E_b|.$$

By using first theorem of graph theory, we get

$$\sum_{(u,v)\in V_{G_a \bigcirc G_b}} deg(u,v) =$$

$$2\times \textit{Number of edges} =$$

$$4\left(\frac{|V_a|(|V_a|-1)}{2} - |E_a|\right)\left(\frac{|V_b|(|V_b|-1)}{2} - |E_b|\right) + 4|E_a||E_b|.$$

Example 6.2 *Consider the graphs G_a and G_b and their modular product $G_a \bigcirc G_b$ given in Example 6.1. Here we have,*

$$\sum_{(u,v)\in V_{G_a \bigcirc G_b}} deg(u,v) = 120$$

and

$$4\left(\frac{|V_a|(|V_a|-1)}{2} - |E_a|\right)\left(\frac{|V_b|(|V_b|-1)}{2} - |E_b|\right) + 4|E_a||E_b| =$$
$$4.[(5.4)/2-4][(4.3)/2-3]+4.4.3 = 120$$

That is,

$$\sum_{(u,v)\in V_{G_a \bigcirc G_b}} deg(u,v) = 4\left(\frac{|V_a|(|V_a|-1)}{2} - |E_a|\right)\left(\frac{|V_b|(|V_b|-1)}{2} - |E_b|\right) + 4|E_a||E_b|.$$

6.4 SOME PROPERTIES OF HOMOMORPHIC PRODUCT OF GRAPHS

Theorem 6.3 *Let $G_a = (V_a, E_a)$ and $G_b = (V_b, E_b)$ be two graphs. Then the homomorphic product of G_a and G_b, which is denoted by $G_a \ltimes G_b$ contains $|V_a||V_b|$ vertices and $(|V_a|\binom{|V_b|}{2} + |E_a|\,[|V_b|(|V_b|-1) - 2|E_b|])$ edges, where $\binom{|V_b|}{2}$ denotes the number of different combinations of vertices in $|V_b|$ taking 2 at a time.*

Proof *The vertex set of $G_a \ltimes G_b$ is $V_a \times V_b$ which contains $|V_a||V_b|$ elements. Therefore total count of vertices in $G_a \ltimes G_b$ is $|V_a||V_b|$. Also, the vertices (v_j, v_k) and (v_l, v_m) in $G_a \ltimes G_b$ are adjacent if and only if*

1. *$v_j = v_l$ or*
2. *v_j and v_l are adjacent in G_a and v_k and v_m are not adjacent in G_b.*

Each edge in $G_a \ltimes G_b$ was created by only one of these two conditions, and neither of them can be true simultaneously. Therefore, we add the number of edges produced by each condition to obtain the overall edge count in $G_a \ltimes G_b$. Think about the first need for adjacency, which is that $v_j = v_l$. Let v be any vertex in G_a. The graph G_b contains $|V_b|$ vertices. We can choose 2 different vertices v' and v'' from G_b in $\binom{|V_b|}{2}$ different ways. Corresponding to each choice we get an edge joining the vertices (v, v') and (v, v'') in $G_a \ltimes G_b$. Like v, there are totally $|V_a|$ vertices in G_a. That is, the first requirement of adjacency results in $|V_a|\binom{|V_b|}{2}$ edges in $G_a \ltimes G_b$. Now think about the second need for adjacency, which is that v_j and v_l are adjacent in G_a but v_k and v_m are not adjacent in G_b. We can select two different vertices v_j and v_l in G_a such that v_j and v_l are adjacent in G_a in $|E_a|$ different ways. Similarly, we can select two different vertices v_k and v_m in G_b such that v_k and v_m are not adjacent in G_b in $\left(\frac{|V_b|(|V_b|-1)}{2} - |E_b|\right)$ different ways. Let v_j and v_l be two vertices in G_a such that v_j and v_l are adjacent in G_a and let v_k and v_m be two vertices in G_b such that v_k and v_m are not adjacent in G_b. From this, we obtain two edges in $G_a \ltimes G_b$, one of which connects the vertices (v_j, v_k) and (v_l, v_m) and the other of which connects the vertices (v_j, v_m) and (v_l, v_k). Hence, totally the second condition for adjacency gives $2|E_a|\left(\frac{|V_b|(|V_b|-1)}{2} - |E_b|\right)$ edges in $G_a \ltimes G_b$. Hence the total count of edges in $G_a \ltimes G_b$ is $|V_a|\binom{|V_b|}{2} + 2|E_a|\left(\frac{|V_b|(|V_b|-1)}{2} - |E_b|\right)$. Therefore total count of edges in $G_a \ltimes G_b$ is

$$|V_a|\binom{|V_b|}{2} + 2|E_a|\left[\frac{|V_b|(|V_b|-1)}{2} - |E_b|\right] = |V_a|\binom{|V_b|}{2} + |E_a|\,[|V_b|(|V_b|-1) - 2|E_b|].$$

Example 6.3 *Let $G_a = (V_a, E_a)$ be a graph having vertex set $V_a = \{v_1, v_2, v_3, v_4, v_5\}$ and edge set $E_a = \{v_1v_3, v_2v_3, v_3v_4, v_4v_5\}$ which is shown in Figure 6.4.*

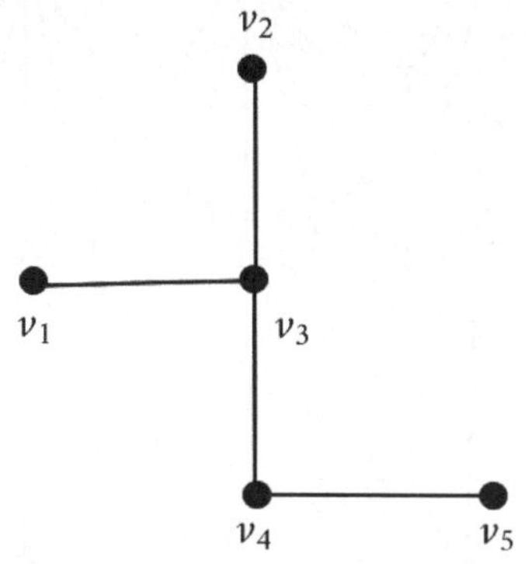

Figure 6.4 Graph $G_a = (V_a, E_a)$.

Consider another graph $G_b = (V_b, E_b)$ having vertex set $V_b = \{u_1, u_2, u_3, u_4\}$ and edge set $E_b = \{u_1u_3, u_3u_4, u_2u_3\}$ which is shown in Figure 6.5.

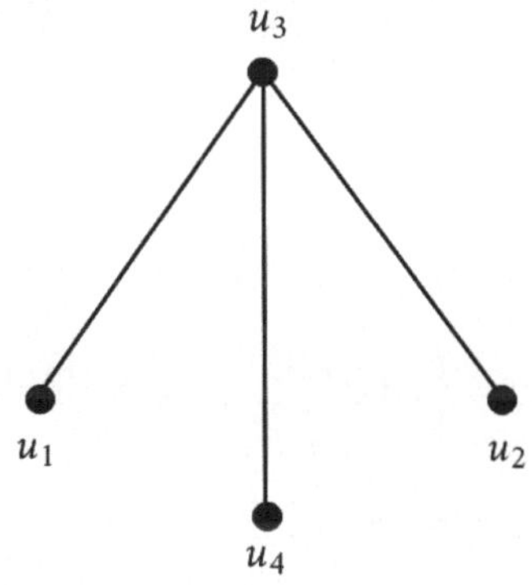

Figure 6.5 Graph $G_b = (V_b, E_b)$.

Then the homomorphic product of these two graphs G_a and G_b is given by $G = G_a \ltimes G_b$ and is shown in Figure 6.6.
Consider the graphs G_a and G_b and their homomorphic product $G_a \ltimes G_b$. Here we have, the overall count of vertices in $G_a \ltimes G_b$ = 20 and

$$|V_a||V_b| = 5 \cdot 4 = 20.$$

That is, the overall count of vertices in $G_a \ltimes G_b = |V_a||V_b|$. Also, the overall count of edges in $G_a \ltimes G_b = 54$ and

$$\left(|V_a|\binom{|V_b|}{2} + |E_a|\,[|V_b|(|V_b|-1) - 2|E_b|]\right) = 5.6 + 4.(4.3 - 2.3) = 54.$$

That is, the overall count of edges in $G_a \ltimes G_b$=

$$\left(|V_a|\binom{|V_b|}{2} + |E_a|\,[|V_b|(|V_b|-1) - 2|E_b|]\right).$$

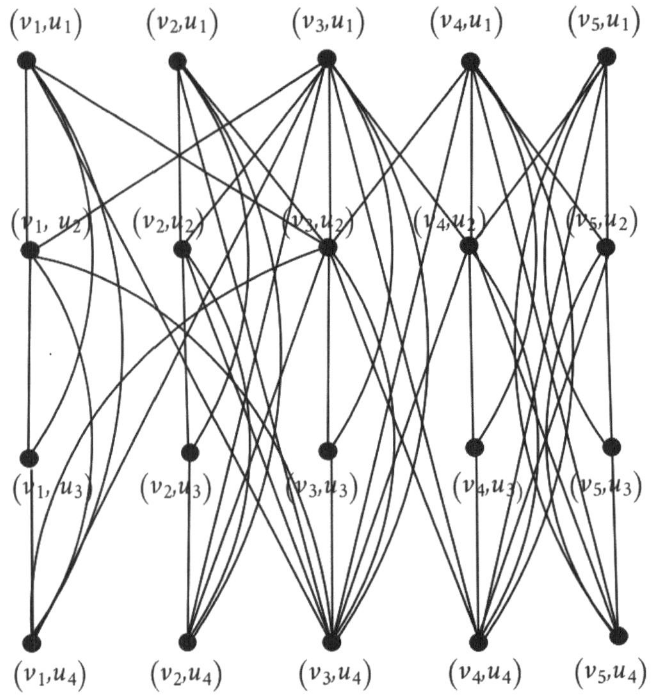

Figure 6.6 $G = G_a \ltimes G_b$.

Theorem 6.4 *Let $G_a = (V_a, E_a)$ and $G_b = (V_b, E_b)$ be two graphs. Then*

$$\sum_{(u,v)\in V_{G_a \ltimes G_b}} deg(u,v) = \left(2|V_a|\binom{|V_b|}{2} + 2|E_a|\left[|V_b|(|V_b|-1) - 2|E_b|\right]\right),$$

where $deg(u,v)$ denotes the degree of the vertex (u,v) in $G_a \ltimes G_b$.

Proof *By Theorem 6.3, we have the overall count of edges in $G_a \ltimes G_b$ is*

$$|V_a|\binom{|V_b|}{2} + |E_a|\left[|V_b|(|V_b|-1) - 2|E_b|\right].$$

By using the first theorem of graph theory, we get

$$\sum_{(u,v)\in V_{G_a \ltimes G_b}} deg(u,v) = 2|V_a|\binom{|V_b|}{2} + 2|E_a|\left[|V_b|(|V_b|-1) - 2|E_b|\right].$$

Hence,

$$\sum_{(u,v)\in V_{G_a \ltimes G_b}} deg(u,v) = \left(2|V_a|\binom{|V_b|}{2} + 2|E_a|\left[|V_b|(|V_b|-1) - 2|E_b|\right]\right).$$

Example 6.4 *Consider the graphs G_a and G_b and their homomorphic product $G_a \ltimes G_b$ given in Example 6.3. Here we have,*

$$\sum_{(u,v)\in V_{G_a \ltimes G_b}} deg(u,v) = 108$$

and

$$\left(2|V_a|\binom{|V_b|}{2}+2|E_a|\left[|V_b|(|V_b|-1)-2|E_b|\right]\right)=2.5.6+2.4.(4.3-2.3)=108.$$

That is,

$$\sum_{(u,v)\in V_{G_a\ltimes G_b}} deg(u,v)=\left(2|V_a|\binom{|V_b|}{2}+2|E_a|\left[|V_b|(|V_b|-1)-2|E_b|\right]\right).$$

6.5 CONCLUSION

Modular product and Homomorphic product are two different graph products in graph theory. In this paper, we investigated some properties of these product operations like the overall vertex and edge counts and the sum of vertex degrees.

1. J. Clark, D. A Holton, *A first look at graph theory*, Allied Publishers Ltd., 1995.
2. G. Chartrand, L. Lesniak, P. Zhang, *Graphs & Digraphs*, CRC Press, 2016.
3. R. Hammack, W. Imrich, S. Klavzar, *Handbook of Product Graphs*, CRC Press, 2011.

7 Results on Transcendental Meromorphic Function of Finite Order on Fixed Points of Difference Polynomials

Priyanka V
Department of Mathematics, School of Engineering, Presidency University, Bangalore, India

Rajeshwari S
Department of Mathematics, Bangalore Institute of Technology, Bangalore, India

Husna V
Department of Mathematics, School of Engineering, Presidency University, Bangalore, India

CONTENTS

In this research paper, the focus revolves around the exploration of $f(z)$, characterized as a transcendental meromorphic function with a finite order $\sigma(f)$, alongside complex constant $c \in \mathbb{C}$. The authors introduce and investigate fixed points concerning the difference polynomials $\Phi(z) = \Delta f(z) - a(f(z))^n$, wherein $\Delta f(z) = f(z+c) - f(z)$. These findings are an extension and expansion of the outcomes previously obtained by Z. Wu and J. Wu.

7.1 INTRODUCTION

In this particular paper, the investigation revolves around exploring the value distribution of difference polynomials of meromorphic functions utilizing the Nevanlinna theory as a framework. To delve into this study, the authors rely on the foundational notations of the Nevanlinna theory, presuming the reader's familiarity with these notations ([1], [2], [3]).
Consider a non-constant meromorphic function $f(z)$ defined across the complex plane $\mathbb{C}$. Throughout our analysis, we will denote the order of $f(z)$ as $\sigma(f)$ and its lower order by $\mu(f)$. The notation $S(r, f)$ will represent any quantity satisfying $S(r, f) = O(T(r, f))$ as r tends toward infinity, potentially excluding a set E possessing finite logarithmic measure.

DOI: 10.1201/9781032634142-7

In the following, a meromorphic function always means meromorphic in the whole complex plane. By $S(r,f)$, we denote any quantity satisfying $S(r,f) = O(T(r,f))$ as $r \to \infty$ possible outside of an exceptional set E with finite logarithmic measure $\int_E dr/r < \infty$. A meromorphic function α is said to be a small function of f if it satisfies $T(r,\alpha) = S(r,f)$.
Let f be a nonconstant meromorphic function. The order of f is defined by

$$\rho(f) = \overline{\lim_{r\to\infty}} \frac{\log^+ T(r,f)}{\log r}.$$

The exponent of convergence of poles of f is defined by

$$\lambda\left(\frac{1}{f}\right) = \overline{\lim_{r\to\infty}} \frac{\log^+ N(r,f)}{\log r}.$$

Let a be a complex number, and let f be a nonconstant meromorphic function. If

$$\overline{\lim_{r\to\infty}} \frac{\log^+ N\left(r,\frac{1}{f-a}\right)}{\log r} < \rho(f),$$

for $\rho(f) > 0$; and $N\left(r,\frac{1}{f-a}\right) = O(\log r)$ for $\rho(f) = 0$, then a is called a Borel exceptional value of f.
Let f and g be two meromorphic functions, and let α be a small functions of both f and g. We say that f and g share α CM (IM) if $f-\alpha$ and $g-\alpha$ have the same zeros counting multiplicities (ignoring multiplicities). $N(r,\alpha)$ is the counting function of common zeros of $f-\alpha$ and $g-\alpha$ with the same multiplicities and the multiplicity is counted. If $N\left(r,\frac{1}{f-\alpha}\right) + N\left(r,\frac{1}{g-\alpha}\right) - 2N(r,\alpha) \leq S(r,f) + S(r,g)$, then we call that f and g share α CM almost. Let $\overline{N}_{(k}(r,f)$ be the counting function for poles of f with multiplicity $\geq k$ where multiplicity is not counted. Set $N_k(r,f) = \overline{N}(r,f) + \overline{N}_{(2}(r,f) + \cdots + \overline{N}_{(k}(r,f)$.
Let $f(z)$ be a function meromorphic in the complex plane $\mathbb{C}$. The order of $f(z)$ is denoted by $\sigma(f)$. For any $a \in \mathbb{C}$, we use the notations $\sigma(f)$ to denote the order of $f(z)$, $\lambda(f,a)$, and $\lambda(1/f)$, respectively, to denote the exponent of convergence of zeros of $f(z) - a$ and poles of $f(z)$. Especially, if $a = 0$, we denote $\lambda(f,0) = \lambda(f)$. A point $z \in \mathbb{C}$ is called as a fixed point of $f(z)$ if $f(z) = z$. There is a considerable number of results on the fixed points for meromorphic functions in the plane, we refer the reader to Chuang and Yang [4]. It follows Chen and Shon [5], we use the notation $\tau(f)$ to denote the exponent of convergence of fixed points of f that is defined as

$$\tau(f) = 1 - \limsup_{r\to\infty} \frac{logN\left(r,\dfrac{1}{f-z}\right)}{logr},$$

Theorem 7.1 *[5] Let f be a transcendental meromorphic function of order of growth $\sigma(f) \leq 1$. Suppose that f satisfies $\lambda(1/f) < \lambda(f) < 1$ or has infinitely many zeros (with $\lambda(f) = 0$) and finitely many poles. Then Δf has infinitely many fixed points and satisfies the exponent of convergence of fixed points $\tau(\Delta f) = \sigma(f)$.*

In 2013, for the existence on the fixed points of differences, Cui and Yang have proved the following theorems.

Theorem 7.2 *[8] Let f be a function transcendental and meromorphic in the plane with the order $\sigma(f) = 1$. If f has finitely many poles and infinitely many zeros with exponent of convergence of zeros $\lambda(f) \neq 1$, then Δf has infinitely many zeros and fixed points.*

Theorem 7.3 *[8] Let f be a function transcendental and meromorphic in the plane with the order $\sigma(f)=1$ max$\{\lambda(f),\lambda(\frac{1}{f})\}=1$. If f has infinitely many zeros, then Δf has infinitely many zeros and fixed points.*

The conditions of Theorems 1.2 and 1.3 imply that $0,\infty$ are Borel exceptional values. If ∞ and $d\in\mathbb{C}$ are Borel exceptional values of f, Chen obtains the following theorem.

Theorem 7.4 *[9] Let f be a finite order meromorphic function such that $\lambda(\frac{1}{f})<\sigma(f)$, and let $c\in\mathbb{C}\{0\}$ be a constant such that $f(z+c)\not\equiv f(z)$. If $f(z)$ has a Borel exceptional value $d\in\mathbb{C}$, then $\tau(\Delta_c f)=\sigma(f)$.*

In 2014, Zhaojun Wu and Hongyan Xu have proved the following.

Theorem 7.5 *[7] Let f be a transcendental meromorphic function of order of growth $\sigma(f)<1$ and $a\in\mathbb{C}$. Suppose that f satisfies $\lambda(1/f)<\sigma(f)$ and $\lambda(f,a)<\sigma(f)$. Then Δf has infinitely many fixed points and satisfies the exponent of convergence of fixed points $\tau(\Delta f)=\sigma(f)$.*

In 2016, Zhang and Chen have obtained the following result.

Theorem 7.6 *[10] Let f be a finite order meromorphic function, and let $c\in\mathbb{C}\{0\}$ be a constant such that $f(z+c)\not\equiv f(z)$. If $f(z)$ has two Borel exceptional values, then $\tau(\Delta_c f)=\sigma(f)$.*

In 2019, Zhaojun Wu and Jia Wu, if the order of f is not a positive integer, have obtained the following result.

Theorem 7.7 *[11] Let f be a transcendental meromorphic function of finite order in the plane. Suppose that $c\in\mathbb{C}\backslash\{0\}$ such that $\Delta_c f\not\equiv 0$. If there is $a\in\mathbb{C}$ with $\delta(a,f)>0$ and $\delta(\infty,f)=1$, then $\Delta_c f$ have infinitely many fixed points and $\tau(\Delta_c f)=\sigma(f)$.*

Theorem 7.8 *[11] Let f be a transcendental meromorphic function of finite order in the plane. Suppose that $c\in\mathbb{C}\backslash\{0\}$ such that $\Delta_c f\not\equiv 0$. If $\delta(\infty,f)=1$, $\delta(0,f)=1$, then*

$$T(r,\Delta_c f)\sim T(r,f)\sim N\left(r,\frac{1}{\Delta_c f-z}\right),$$

as $r\longrightarrow\infty$, $r\notin E$, where E is a possible exception set of r with finite logarithmic measure.

Let $f(z)$ be a meromorphic function in the complex plane $\mathbb{C}$ and $a\in\mathbb{C}_\infty=\mathbb{C}\cup\{\infty\}$. Nevanlinna's deficiency of f with respect to a is defined by

$$\delta(a,f)=1-\limsup_{r\longrightarrow\infty}\frac{N\left(r,\frac{1}{f-a}\right)}{T(r,f)},$$

If $a=\infty$, then one should replace $N\left(r,\frac{1}{f-a}\right)$ in the above formula by $N(r,f)$. If $\delta(a,f)>0$, then a is called a Nevanlinna deficiency value of f.

Let $f(z)$ be a transcendental meromorphic function and $a\in\mathbb{C}\backslash\{0\}$, and $c\in\mathbb{C}\backslash\{0\}$ be complex constants satisfying. We have investigated the fixed point of a meromorphic function with the difference polynomials

$$\Phi(z)=\Delta f(z)-a(f(z))^n, \tag{7.1}$$

where $\Delta f(z)=f(z+c)-f(z)$.

Definition 7.1 *[6] Let p be a positive integer and $a \in \mathbb{C} \cup \{\infty\}$.*

(i) $N_{(p}(r,1/(f-a))$ $(\overline{N}_{(p}(r,1/(f-a)))$ denotes the counting function (reduced counting function) of those a-points of f whose multiplicities are not less than p.
(ii) $N_{p)}(r,1/(f-a))$ $(\overline{N}_{p)}(r,1/(f-a)))$ denotes the counting function (reduced counting function) of those a-points of f whose multiplicities are not greater than p.

Definition 7.2 *[6] Let f and g be two non-constant meromorphic functions such that f and g share the value a IM. Let z_0 be a a-point of f and g with respective multiplicities p and q. We denote by $\overline{N}_L(r,1/(f-a))$ the counting function of those a-points of f and g where $p>q$, by $N_E^{1)}(r,1/(f-a))$ the counting function of those a-points of f and g where $p=q=1$ and by $\overline{N}_E^{(2}(r,1/(f-a))$ the counting function of those a-points of f and g where $p=q\geq 2$, each point in these counting functions is counted only once. Similarly, one can define $\overline{N}_L(r,1/(f-a)), N_E^{1)}(r,1/(f-a)), \overline{N}_E^{(2}(r,1/(f-a))$.*

Definition 7.3 *[6] Let f, g share a value a IM. We denote by $\overline{N}_*(r,a;f,g)$ the reduced counting function of those a-points of f whose multiplicities differ from that of the corresponding a-points of g.*
It is easy to see that $\overline{N}_*(r,a;f,g) \equiv \overline{N}_*(r,a;g,f)$ and $\overline{N}_*(r,a;f,g) = \overline{N}_L(r,a;f) + \overline{N}_L(r,a;g)$.

Definition 7.4 *[6] For $a \in \mathbb{C} \cup \{\infty\}$ and a positive integer p we denote by $N_p(r,1/(f-a))$ $\overline{N}(r,1/(f-a)) + \overline{N}_{(2}(r,1/(f-a)) + \cdots + \overline{N}_{(p}(r,1/(f-a))$. It is easy to see that $N_1(r,1/(f-a)) = \overline{N}(r,1/(f-a))$.*

Remark 7.1 *If $a=0$, then $\Phi(z)$ becomes the forward difference $\Delta_c f$, i.e.*

$$\Phi(z) = f(z+c) - f(z) = \Delta_c f(z).$$

7.2 SOME LEMMAS

In this section, we present some lemmas which will be needed later on to prove Main results .

Lemma 7.1 *[12] Let $f(z)$ be a transcendental meromorphic function of finite order, then*

$$m\left(r, \frac{f(z+c)}{f}\right) = S(r,f).$$

Lemma 7.2 *[18] Let $f(z)$ be a finite order meromorphic function, then, for each $k \in \mathbb{N}$, $\sigma(\Delta_c^k f) \leq \sigma(f)$.*

Lemma 7.3 *[13] Let f be a transcendental meromorphic function of finite order. Then for any positive integer n, we have*

$$m\left(r, \frac{\Delta_c^n f(z)}{f(z)}\right) = S(r,f).$$

Lemma 7.4 *[14][15] Let f be a transcendental meromorphic function of finite order. Then*

$$N(r,f(z+c)) = N(r,f) + S(r,f),$$

$$T(r,f(z+c)) = T(r,f) + S(r,f),$$

where $S(r,f) = O(T(r,f))$ $(r \longrightarrow \infty)$, possibly outside a set E of r with finite logarithmic measure.

Lemma 7.5 *[16] [17] Suppose that $f(z)$ is a transcendental meromorphic function in the complex plane and $P(z) = a_0 z^n + a_1 z^{n-1} + \cdots + a_n$, where $a_0 (\not\equiv 0)$, $a_1, \ldots, a_n$ are constants. Then*

$$T(r, P(f)) = nT(r, f) + S(r, f).$$

Lemma 7.6 *[16] [20] Let $F(r)$ and $G(r)$ be monotone increasing function such that $F(r) \le G(r)$ outside of exceptional set E that is of finite logarithmic measure. Then for any $\alpha > 0$, there exists $r_0 > 1$ such that $F(r) \le G(\alpha r)$ for all $r > r_0$.*

Lemma 7.7 *Let $f(z)$ be a transcendental meromorphic function of finite order $\sigma(f)$ in the complex plan. let $c \in \mathbb{C}$ be complex constants satisfying such that $\Phi(z) \not\equiv 0$, and $\delta(0, f) > 0$. Then $\Phi(z)$ a transcendental and meromorphic function of finite order.*

Proof *Since $\delta(0, f) > 0$, from Lemma 7.2, we know that $\sigma(\Phi(z)) \le \sigma(f) < +\infty$. If $\Phi(z)$ is a transcendental meromorphic function. Suppose that $\Phi(z)$ is not a transcendental meromorphic function. Then there is a rational $Q(z)$ such that $Q(z)\Phi(z) \equiv 1$, i.e.,*

$$\begin{aligned}\frac{1}{f^{n+1}} &\equiv Q(z)\frac{\Phi(z)}{f^{n+1}} \\ &\equiv Q(z)\left(\left(\frac{f(z+c)}{f}\right) - \left(\frac{f(z)}{f}\right) - a\left(\frac{f(z)}{f}\right)^n\right).\end{aligned}$$

Apply Lemma 7.1 and note that $f(z)$ is transcendental, we can get

$$m\left(r, \frac{1}{f^{n+1}}\right) \le m(r, Q(z)) + m\left(r, \frac{\Phi(z)}{f^{n+1}}\right) = S(r, f).$$

Therefore

$$\begin{aligned}m\left(r, \frac{1}{f^{n+1}}\right) + N\left(r, \frac{1}{f^{n+1}}\right) &\le N\left(r, \frac{1}{f^{n+1}}\right) + S(r, f) \\ &\le (n+1)N\left(r, \frac{1}{f}\right) + S(r, f).\end{aligned}$$

Apply Lemma 7.5 and the first fundamental theorem of Nevanlinna theory, we can get

$$(n+1)T(r, f) \le (n+1)N\left(r, \frac{1}{f}\right) + S(r, f).$$

This contradicts with $\delta(0, f) > 0$. Thus $\Phi(z)$ is a transcendental and meromorphic function of finite order.

Lemma 7.8 *Let $f(z)$ be a transcendental meromorphic function of finite order $\sigma(f)$ in the complex plane. Let $c \in \mathbb{C}$ be complex constants satisfying $\Phi(z) \not\equiv 0$ and $\delta(0, f) > 0$, then*

$$(n+1)T(r, f) \le (n+1)N\left(r, \frac{1}{f}\right) + 2(n+1)N(r, f) + N\left(r, \frac{1}{\Phi(z) - z}\right) + S(r, f).$$

Proof *By Lemma 7.7, we know that $\Phi(z)$ is a transcendental meromorphic function, then there is $\eta \in \mathbb{C}\backslash\{0\}$ such that $z\Delta_\eta \Phi(z) - \Phi(z) \not\equiv 0$.*
Noticing

$$\frac{1}{f^{n+1}} = \frac{\Phi(z)}{z f^{n+1}} - \frac{z\Delta_\eta \Phi(z) - \Phi(z)}{z f^{n+1}} \frac{\Phi(z) - z}{z\Delta_\eta \Phi(z) - \Phi(z)}, \tag{7.2}$$

then

$$m\left(r,\frac{1}{f^{n+1}}\right) \leq 2m\left(r,\frac{\Phi(z)}{f^{n+1}}\right)+m\left(r,\frac{\Delta_\eta\Phi(z)}{f^{n+1}}\right)+m\left(r,\frac{\Phi(z)-z}{z\Delta_\eta\Phi(z)-\Phi(z)}\right)+O(logr). \tag{7.3}$$

Applying the first fundamental theorem, we get

$$m\left(r,\frac{1}{f^{n+1}}\right)=(n+1)T(r,f)-(n+1)N\left(r,\frac{1}{f}\right)+O(1). \tag{7.4}$$

$$\begin{aligned} m\left(r,\frac{\Phi(z)-z}{z\Delta_\eta\Phi(z)-\Phi(z)}\right) =& m\left(r,\frac{z\Delta_\eta\Phi(z)-\Phi(z)}{\Phi(z)-z}\right)+N\left(r,\frac{z\Delta_\eta\Phi(z)-\Phi(z)}{\Phi(z)-z}\right) \\ &-N\left(r,\frac{\Phi(z)-z}{z\Delta_\eta\Phi(z)-\Phi(z)}\right)+O(1) \\ \leq& m\left(r,\frac{z\Delta_\eta\Phi(z)-\Phi(z)}{\Phi(z)-z}\right)+N\left(r,\frac{z\Delta_\eta\Phi(z)-\Phi(z)}{\Phi(z)-z}\right)+O(1). \end{aligned} \tag{7.5}$$

Combining (2)-(4) we have

$$\begin{aligned} (n+1)T(r,f) \leq& (n+1)N\left(r,\frac{1}{f}\right)+2m\left(r,\frac{\Phi(z)}{f^{n+1}}\right)+m\left(r,\frac{\Delta_\eta\Phi(z)}{f^{n+1}}\right) \\ &+m\left(r,\frac{z\Delta_\eta\Phi(z)-\Phi(z)}{\Phi(z)-z}\right) \\ &+N\left(r,\frac{z\Delta_\eta\Phi(z)-\Phi(z)}{\Phi(z)-z}\right)+O(logr) \\ \leq& (n+1)N\left(r,\frac{1}{f}\right)+N\left(r,\frac{1}{\Phi(z)-z}\right)+N(r,z\Delta_\eta\Phi(z)-\Phi(z)) \\ &+2m\left(r,\frac{\Phi(z)}{f^{n+1}}\right)+m\left(r,\frac{\Delta_\eta\Phi(z)}{f^{n+1}}\right)+m\left(r,\frac{z\Delta_\eta\Phi(z)-\Phi(z)}{\Phi(z)-z}\right)+O(logr). \end{aligned} \tag{7.6}$$

Since

$$\begin{aligned} \Delta_\eta\Phi(z) &= \Delta_\eta(f(z+c)-f(z)-a(f(z))^n \\ &= f(z+c+\eta)-f(z+\eta)-a((f(z+\eta))^n-(f(z))^n)-f(z+c)+f(z) \\ \Delta_\eta(\Phi(z)-z) &= f(z+c+\eta)-f(z+\eta)-a((f(z+\eta))^n-(f(z))^n)-f(z+c)+f(z)-(z+\eta), \end{aligned}$$

then, we can get

$$\begin{aligned} z\Delta_\eta\Phi(z)-\Phi(z) =& z(f(z+c+\eta)-f(z+\eta))-az((f(z+\eta))^n-2(f(z))^n) \\ &+a(f(z))^n-(z+1)(f(z+c)-f(z)). \\ z\Delta_\eta(\Phi(z)-z)-(\Phi(z)-z) =& z(f(z+c+\eta)-f(z+\eta))-az((f(z+\eta))^n-2(f(z))^n) \\ &+a(f(z))^n-(z+1)(f(z+c)-f(z))-z(z+\eta-1) \end{aligned}$$

Therefore,

$$\begin{aligned} \frac{z\Delta_\eta\Phi(z)-\Phi(z)}{\Phi(z)-z} &= \frac{z\Delta_\eta(\Phi(z)-z)-(\Phi(z)-z)}{\Phi(z)-z} \\ &= \frac{z\Delta_\eta(\Phi(z)-z)}{\Phi(z)-z}-1 \end{aligned} \tag{7.7}$$

$$\begin{aligned} N(r,z\Delta_\eta\Phi(z)-\Phi(z)) \leq& N(r,f(z+c+\eta)-f(z+\eta)) \\ &+N(r,(f(z+\eta))^n-2(f(z))^n) \\ &+N(r,(f(z))^n)+N(r,f(z+c)-f(z)). \end{aligned} \tag{7.8}$$

Thus from Lemma 7.4 and (7.7), we deduce

$$N(r, z\Delta_\eta\Phi(z) - \Phi(z)) \leq 2(n+1)N(r, f(z)) + O(logr). \tag{7.9}$$

By Lemmas 7.2 and 7.7 we know that $\Phi(z) - z$ is a transcendental meromorphic function of finite order. It follows from Lemma 7.3 and (7.8), we get

$$\begin{aligned} m\left(r, \frac{\Phi(z)}{f^{n+1}}\right) &= S(r,f) \\ m\left(r, \frac{\Delta_\eta\Phi(z)}{f^{n+1}}\right) &= S(r,f) \\ m\left(r, \frac{z\Delta_\eta\Phi(z) - \Phi(z)}{\Phi(z) - z}\right) &= S(r,f). \end{aligned} \tag{7.10}$$

From (7.5), and (7.8), and (7.9), we have

$$(n+1)T(r,f) \leq (n+1)N\left(r, \frac{1}{f}\right) + 2(n+1)N(r,f) + N\left(r, \frac{1}{\Phi(z) - z}\right) + S(r,f). \tag{7.11}$$

7.3 PROOF OF MAIN RESULTS

Theorem 7.9 *Let $f(z)$ be a transcendental meromorphic function of finite order $\sigma(f)$ in the complex plane. let $c \in \mathbb{C}$ be complex constants satisfying such that $\Phi(z) \not\equiv 0$, If there is $a \in \mathbb{C}$ with f is satisfies $\delta(\infty, f) = 1$, and a is a Nevanlinna deficiency value of f, then $\Phi(z)$ have infinitely many fixed points and $\tau(\Phi(z)) = \sigma(f)$.*

Proof *Denoting $g = f - a$ by (7.10) we derive,*

$$\begin{aligned} (n+1)T(r,f) \leq& (n+1)T(r,g) + O(1) \\ \leq& (n+1)N\left(r, \frac{1}{g}\right) + 2(n+1)N(r,g) + N\left(r, \frac{1}{\Phi(z) - z}\right) + S(r,g) \\ =& (n+1)N\left(r, \frac{1}{f-a}\right) + 2(n+1)N(r,f) + N\left(r, \frac{1}{\Phi(z) - z}\right) + S(r,f). \end{aligned} \tag{7.12}$$

Since $\delta(0, f) > 0$ and $\delta(\infty, f) = 1$,
there is a positive number $\theta < 1$, such that

$$N\left(r, \frac{1}{f-a}\right) < \theta T(r,f) \tag{7.13}$$

$$N(r,f) = O(1) + T(r,f). \tag{7.14}$$

If $\Phi(z)$ has only a finite number of fixed points, then from (7.12)-(7.14), we would have

$$(n+1)(1 - O(1) - \theta)T(r,f) \leq N\left(r, \frac{1}{\Phi(z) - z}\right), r \notin E, r \longrightarrow \infty, \tag{7.15}$$

where E is a possible exceptional set with finite logarithmic measure. Noticing f is transcendental, applying Lemma 7.6 and (7.15), we can get that $\Phi(z)$ assumes has infinitely many fixed points and $\tau(\Phi(z)) = \sigma(f)$.

Theorem 7.10 *Let $f(z)$ be a transcendental meromorphic function of finite order $\sigma(f)$ in the complex plane. $c \in \mathbb{C}$ be complex constants satisfying such that $\Phi(z) \not\equiv 0$, If there is $a \in \mathbb{C}$ with f is satisfies $\delta(\infty, f) = 1$ and $\delta(0, f) = 1$, then*

$$T(r, \Phi(z)) \sim (n+1)T(r, f) \sim N\left(r, \frac{1}{\Phi(z) - z}\right),$$

as $r \longrightarrow \infty$, $r \notin E$, where E is a possible exception set of r with finite logarithmic measure.

Proof *Since $\delta(0, f) = 1$ and $\delta(\infty, f) = 1$,*

$$N\left(r, \frac{1}{f}\right) = S(r, f) \tag{7.16}$$

$$N(r, f) = S(r, f) \tag{7.17}$$

Since

$$T(r, \Phi(z)) \le T(r, f(z+c) - f(z)) + T(r, (f(z))^n). \tag{7.18}$$

Using Lemma 7.4, we can derive from (7.18) that

$$T(r, \Phi(z)) \le (1+n)T(r, f) + S(r, f). \tag{7.19}$$

From (7.16)-(7.19), we have

$$\begin{aligned} (n+1)T(r, f) &\le N\left(r, \frac{1}{\Phi(z) - z}\right) + S(r, f) \\ &\le T(r, \Phi(z)) + S(r, f) \\ &\le (n+1)T(r, f) + S(r, f). \end{aligned} \tag{7.20}$$

Since f is transcendental, (7.20) means that $\Phi(z)$ assumes has infinitely many fixed points and

$$T(r, \Phi(z)) \sim (n+1)T(r, f) \sim N\left(r, \frac{1}{\Phi(z) - z}\right),$$

as $r \notin E, r \longrightarrow \infty$, where E is a possible exception set of r with finite logarithmic measure.

Funding Agency. There is no funding agency.
Conflict of Interest. There is no conflict of interest. This article doesn't contain any studies with human participants and animals performed by any of the author.

REFERENCES

1. Walter Kurt Hayman. *Meromorphic functions*. Oxford Clarendon Press, 1964.
2. Chung-Chun Yang and Hong-Xun Yi. *Uniqueness theory of meromorphic functions*. Springer Science & Business Media, 2004.
3. Jianhua Zheng. *Singular values of meromorphic functions*. In: Value Distribution of Meromorphic Functions. Springer, 2010, pp. 229–266.
4. CT Chuang and CC Yang. *Theory of Fix Points and Factorization of Mero-Morphic Functions*. Mathematical Monogragh Series, Peking University Press, 1986.
5. Zong-Xuan Chen and Kwang Ho Shon. *Properties of differences of meromorphic functions*. Czechoslovak Mathematical Journal, 61(1), 2011, pp. 213–224.
6. I Lahiri. *Milloux theorem, deficiency and fix-points for vector-valued meromorphic functions*. The Journal of the Indian Mathematical Society, 59(1-4), 1993, pp. 45–60.

7. Zhaojun Wu and Hongyan Xu. *Fixed points of difference operator of meromorphic functions*. The Scientific World Journal, 2014, 2014, Hindawi.
8. Cui Weiwei and Yang Lianzhong. *Zeros and fixed points of difference operators of meromorphic functions*. Acta Mathematica Scientia, 33(3), 2013, pp. 773–780.
9. Zong-Xuan Chen. *Fixed points of meromorphic functions and of their differences and shifts*. In: Annales Polonici Mathematici, vol. 2, no. 109, 2013, pp. 153–163.
10. Ran Ran Zhang and Zong Xuan Chen. *Fixed points of meromorphic functions and of their differences, divided differences and shifts*. Acta Mathematica Sinica, English Series, 32(10), 2016, pp. 1189–1202.
11. Zhaojun Wu and Jia Wu. *Fixed points of differences of meromorphic functions*. Advances in Difference Equations, 2019, 2019, pp. 1–8.
12. RG Halburd and RJ Korhonen. *Meromorphic solutions of difference equations, integrability and the discrete Painleve equations*. Journal of Physics A: Mathematical and Theoretical, 40(6), 2007, pp. R1.
13. R G Halburd and R J Korhonen. *Nevanlinna theory for the difference operator*. Ann. Acad. Sci. Fenn. Math., 31, 2006, pp. 463-478.
14. Yik-Man Chiang and Shao-Ji Feng. *On the Nevanlinna characteristic of $f(z+\eta)$ and difference equations in the complex plane*. The Ramanujan Journal, 16(1), 2008, pp. 105–129.
15. R. R. Zheng and Z. X. Chen. *On value distribution of difference polynomials of meromorphic functions*. Sci. Sin. Math, 42(11), 2012, pp. 1115-1130.
16. ZongXuan Chen. *On growth, zeros and poles of meromorphic solutions of linear and nonlinear difference equations*. Science China Mathematics, 54(10), 2011, pp. 2123–2133.
17. G Gundersen. *Finite order solutions of second order linear differential equations*. Trans Ame Math Soc, 305, 1998, Springer.
18. Yik-Man Chiang and Shao-Ji Feng. *On the growth of logarithmic differences, difference quotients and logarithmic derivatives of meromorphic functions*. Transactions of the American Mathematical Society, 361(7), 2009, pp. 3767–3791.
19. Xiu-Min Zheng and Zong-Xuan Chen. *On the value distribution of some difference polynomials*. Journal of Mathematical Analysis and Applications, 397(2), 2013, pp. 814–821.
20. Gary G Gundersen. *Finite order solutions of second order linear differential equations*. Transactions of the American Mathematical Society, 305(1), 1988, pp. 415–429.
21. V Priyanka, S Rajeshwari, and V Husna. *Results on certain differential polynomials of L-functions sharing a finite value*.
22. V Priyanka, S Rajeshwari, and V Husna. *Uniqueness problems for difference polynomials sharing a non-zero polynomial of certain degree with finite weight*. J. Math. Anal. Appl, 19(2), 2022.
23. V Priyanka, S Rajeshwari, and V Husna. *Uniqueness of differential-difference polynomials of meromorphic functions sharing shift polynomial and small function IM*. Journal— MESA, 14(4), 2023, pp. 1125–1138.
24. V Priyanka, S Rajeshwari, and V Husna. *Generalization on value distribution of L-functions*. Nonlinear Studies, 30(1), 2023.

8 On the Construction of Quasigroups by Planar Mappings over Finite Fields

Tajender Kumar
Department of Mathematics,
Kalindi College, University of Delhi, India

CONTENTS

We construct quasigroups by using planar functions over finite fields. Further we evaluate the associativity index of quasigroups for particular type of polynomials over finite fields.

8.1 INTRODUCTION

A pair $(Q,*)$ is called a quasigroup if the operation $*$ is closed on Q and the equations

$$\begin{aligned} x*a &= b, \\ a*y &= b. \end{aligned} \tag{8.1}$$

have a unique solution for every $a,\ b \in Q$. Hence the multiplication table of each quasigroup is equivalent to a Latin square. A triple (x,y,z) where x, y and $z \in Q$, which satisfies $(x*y)*z = x*(y*z)$, is called associative triple in $(Q,*)$. The algebra of quasigroups were extensively studied by Drapal [8, 9], Norton [17], Ježek and Kepka [11]. There are numerous applications of quasigroups in cryptography; see e.g. [12, 15, 19], for getting the design of a quasigroup based hash function [10, 20].

Let $(Q,*)$ be a quasigroup and let $Q^\star$ be a set of all finite strings over Q by the elements $m_i \in Q$, $1 \le i \le k$. Let $(m_1m_2...m_k)$ be the finite string and the hash function [10] is defined as:

$$H : Q \times Q^\star \to Q$$
$$H(a,m_1m_2...m_k) = a*(m_1*(m_2*...*(m_{k-1}*m_k))),$$

where $a \in Q$ is fixed. We can also write $H(a,m_1m_2...m_k) = H_a(m_1,m_2,...,m_k)$ for simplicity.

A hash function $h : X \to Y$ is said to be collision resistant if it computationally infeasible to find $x,x' \in X$ such that $x' \neq x$ and $h(x') = h(x)$. Now, we check if our designed hash function is collision

DOI: 10.1201/9781032634142-8

resistant or not. If (m_{k-1}, x, y) satisfies associativity axiom in Q and $m_k = x * y$, then we choose $(m_1, m_2, m_3, ..., m_{k-1}, m_k)$ and $(m_1, m_2, m_3, ..., m_{k-1} * x, y)$ two arbitrary finite strings over Q, so that

$$\begin{aligned} H_a(m_1, m_2, m_3, ..., m_{k-1}, m_k) &= a * (m_1 * (m_2 * (m_3 * (... * (m_{k-1} * m_k))))) \\ &= a * (m_1 * (m_2 * (m_3 * (... * (m_{k-1} * (x * y)))))) \\ &= a * (m_1 * (m_2 * (m_3 * (... * ((m_{k-1} * x) * y))))) \\ &= H_a(m_1, m_2, m_3, ..., m_{k-1} * x, y) \end{aligned}$$

or

$$H_a(m_1, m_2, m_3, ..., m_{k-1}, m_k) = H_a(m_1, m_2, m_3, ..., m_{k-1} * x, y). \tag{8.2}$$

Hence the messages $(m_1 m_2 m_3 ... m_{k-1} m_k)$ and $(m_1 m_2 m_3 ... m_{k-1} * xy)$ in Q such that

$$(m_1, m_2, m_3, ..., m_{k-1}, m_k) \neq (m_1, m_2, m_3, ..., m_{k-1} * x, y)$$

but both have the same hash value. Here we see that the problem of collision depends on associative triples in Q. If we have less number of associative triples in Q then our hash function is more secure.

Grošek and Horak [10] state that upper bound for associative triples of any quasigroup Q is less than or equal to $2 \cdot |Q|^2$. Although Kotzig and Reischer [13] constructed infinite class of quasigroups Q with associative triples less than $|Q|^2$, where $|Q|$ is size of the set Q. Therefore, our concern in this paper is to evaluate the counts of associative triples of proposed quasigroups. Ježek and Kepka [11, Section 2] determined bounds on the number of associative triples in the finite quasigroups Q. Drapal and Valent [9] established new upper bound for associativity index of quasigroups of a given order. The upper bound is defined by an average number of associative triples among principal isotopes of a quasigroup of given order. For the permutations $\alpha, \beta \in S_Q$, let $Q_{\alpha,\beta}$ be the principal isotope $x * y = \alpha(x)\beta(y)$. This average number of associative triples is $n^2(1+\varepsilon), \varepsilon^{-1} = n-1$, $|Q| = n$. If G is a group and α is an orthomorphism, then $a(Q_{\alpha,\beta}) = n^2$ for every $\beta \in S_Q$. The known best bound is so far:

Corollary 8.1 [9, Corollary 2.11] *If Q is a commutative quasigroup of order n that is isotopic to a commutative group, then $a(Q) \geq n^2$ if $n \not\equiv 2 \mod 4$, and $a(Q) \geq n^2 + 2n$ if $n \equiv 2 \mod 4$. Both estimates are sharp, for every $n \geq 1$.*

Let $\mathbb{F}_q$ denote the Galois field of order $q = p^n$.

For simplicity, we denote by $\mathbb{F}_q^*$ the set $\mathbb{F}_q \setminus \{0\}$. The details of the paper are as follows: In Section 8.2 we define the lower and upper bound of associative index for quasigroups. In Section 8.3 we define a new construction of quasigroups via permutations and planar functions over $\mathbb{F}_q$. We see that the algebraic properties of this construction extensively depend on the using permutations and planar functions. Therefore we investigate the counts of associative triples for different permutations and planar functions over $\mathbb{F}_q$.

8.2 ASSOCIATIVE TRIPLES IN QUASIGROUP

Quasigroups are non-associative algebraic structure. In a quasi group $(Q, *)$, let $A(Q)$ and $B(Q)$ be defined as:

$$\begin{aligned} A(Q) &= \{(x, y, z) \in Q^3 : (x * y) * z = x * (y * z)\}, \\ B(Q) &= Q^3 - A(Q). \end{aligned}$$

Let $a(Q) = |A(Q)|$ and $b(Q) = |B(Q)|$. If Q is finite with $n = |Q|$ then

$$a(Q) + b(Q) = n^3.$$

There exists a unique left identity a_l and a unique right identity a_r , $for\ all\ a \in Q$. Thus, we get

$$a_l * a = a,$$
$$a * a_r = a.$$

Hence the triple (a_l, a, a_r) is associative for each $a \in Q$ and the lower bound for $a(Q)$ is $a(Q) \geq n$. Therefore, $n \leq a(Q) \leq n^3$.

A quasigroup is said to be di-associative if in $A(Q)$ above, exactly two elements are distinct and mono-associative when all three elements x, y, and z are equal, respectively. The following theorems are proven by Norton [17].

Theorem 8.1 *Let $(Q,*)$ be a quasigroup for which both Q and $Q^2 = \{q*q : for\ all q \in Q\}$ contain a "sufficient number" of elements, then Q is di-associative.*

Theorem 8.2 *If a quasigroup $(Q,*)$ satisfies the constraint $x*(x*y) = (x*x)*y$ when $x \neq y \in Q$, then Q is mono-associative.*

For any quasigroup, we can say that

$$\text{Associativity (tri-associativity)} \Longrightarrow \text{di-associativity} \Longrightarrow \text{mono-associativity.}$$

Our target is to reduce the upper bound of $a(Q)$ and find out the exact number of $a(Q)$ for given quasigroup $(Q,*)$ of order n.

8.3 CONSTRUCTION OF QUASIGROUPS BY PLANAR MAPPINGS OVER FINITE FIELDS

The derivative of any function $f(x) \in \mathbb{F}_q[x]$ at $a \in \mathbb{F}_q[x]$ is defined as follows

Definition 8.1 *For a given polynomial $f(x) \in \mathbb{F}_q[x]$ its derivative at $a \in \mathbb{F}_q^*$ is defined as*

$$D_a f(x) = f(x+a) - f(x)$$

where clearly $a = 0$ results in a trivial annihilation.

In contrast to the standard notion of derivative, which is for instance useful for determination of multiple roots of f and which coincides to the derivation of polynomials over real numbers, this notion of derivatives is of great importance in cryptography and is directly related to differential properties of the mappings used in the substitution boxes.

Indeed, when $p = 2$, the differential properties of f (that reflects the resistance to differential cryptanalysis [1]) are characterized by the number of solutions of

$$f(x+a) + f(x) = b$$

for any $a \in \mathbb{F}_q^*$ and $b \in \mathbb{F}_q$. On the other hand, for fields of odd prime characteristic $p > 2$, if $f(x+a) - f(x)$ is a permutation for any non zero a then f is called a planar function [5, 6, 7].

Definition 8.2 *A mapping $f : \mathbb{F}_q \to \mathbb{F}_q$ is called planar if its derivative $D_a f(x)$ is a permutation of $\mathbb{F}_q$, for every fixed $a \in \mathbb{F}_q^*$.*

Notice that f itself cannot be a permutation, and it is known that if f is considered as $f:\mathbb{F}_p^n \to \mathbb{F}_p^n$ then all nonzero linear combinations of its coordinate functions f_i (representing $f=(f_1,\ldots,f_n)$) are bent functions from $\mathbb{F}_p^n$ to $\mathbb{F}_p$. For above definitions and details we refer to [16].
A linearized polynomial $L:\mathbb{F}_q \to \mathbb{F}_q$

$$L(x)=\sum_{i=0}^{m-1} b_i x^{p^i}$$

and it induces a linear operator on the vector space $\mathbb{F}_q$ over its prime subfield. An affine polynomial is of the form $A(x)=L(x)+c$ for some constant $c\in\mathbb{F}_q$ and a linearized polynomial $L(x)$.

A polynomial $f(x)\in\mathbb{F}_q[x]$ is called quadratic if it is Dembowski–Ostrom (DO) type of polynomial, that is, a polynomial of the form $\sum_{0\leq k,j<n} a_{k,j}x^{p^k+p^j}$, $a_{k,j}\in\mathbb{F}_q$ added an affine polynomial $A(x)$. Almost all known planar mappings are of DO type with a single exception being the power function $f(x)=x^{\frac{3^t+1}{2}}$ over $\mathbb{F}_{3^n}$, where t is odd and $\gcd(t,n)=1$, see [5]. Note also that the case of power functions $f(x)=x^d$ is completely resolved over $\mathbb{F}_{p^2}$, that is, f is planar if and only if $d=2$ or $d=2p$.

Theorem 8.3 [14, Theorem 7.8]
(i) Every linear polynomial over $\mathbb{F}_q$ is a permutation polynomial of $\mathbb{F}_q$.
(ii) The monomial x^n is a permutation polynomial of $\mathbb{F}_q$ if and only if $\gcd(n,q-1)=1$.

Sade [18] proposed a method of constructing quasigroups from complete mapping permutations. The following proposition describes that method.

Proposition 8.1 ([18], Theorem 14) *Let $(Q,+)$ be an admissible group with complete mapping θ. Then, $*:Q\times Q\to Q$ is defined as:*

$$x*y=\theta(x-y)+y$$

where $x,y\in Q$. Then $(Q,*)$ is a quasigroup.

It follows immediately that.

Theorem 8.4 *Suppose $\mathbb{F}_{p^n}=Q$ is the extension field of degree n over $\mathbb{F}_p$. Let f and g are permutation and planar mapping on $\mathbb{F}_{p^n}$ respectively. For $a\in\mathbb{F}_q^*$, it is defined as*

$$x*y=f(x)+D_ag(y) \tag{8.3}$$

for all $x,y\in Q$. Then $(Q,*)$ is a quasigroup.

Proof *For any $a',b'\in Q$, consider the equation*

$$\begin{aligned} a'*x&=b' \\ f(a')+D_ag(x)&=b' \\ D_ag(x)&=f(a')+b' \end{aligned}$$

Since g is planar mapping over $\mathbb{F}_q$. Then we have:

$$\begin{aligned} g(x+a)-g(x)&=f(a')+b' \\ g'(x)&=f(a')+b', \quad \text{where } g'(x)=g(x+a)-g(x) \\ \text{i.e.,}\quad x&=g'^{-1}(f(a')+b') \end{aligned}$$

Similarly for another equation

$$\begin{aligned} y*a' &= b' \\ f(y)+D_a g(a') &= b' \\ f(y) &= d_a g(a')+b' \end{aligned}$$

Since f is permutation over $\mathbb{F}_q$. Then we have:

$$y = f^{-1}(d_a g(a')+b')$$

Therefore both equations have unique solution for every $a', b' \in Q$. Thus $(Q,*)$ is a quasigroup.

8.3.1 COUNTING THE NUMBER OF ASSOCIATIVE TRIPLES

Now we check the associative condition on the above quasigroup.

$$(x*y)*z = x*(y*z) \tag{8.4}$$

Taking left hand side $(x*y)*z$, first go with this part $x*y$, then we get

$$\begin{aligned} (x*y)*z &= (f(x)+D_a g(y))*z \\ &= f(f(x)+D_a g(y))+D_a g(z) \end{aligned} \tag{8.5}$$

Similarly, taking right hand side $x*(y*z)$, first go with this part $y*z$, then we get

$$\begin{aligned} x*(y*z) &= x*(f(y)+D_a g(z)) \\ &= f(x)+D_a g(f(y)+D_a g(z)) \end{aligned} \tag{8.6}$$

By Equations (5) and (6), we have

$$f(f(x)+D_a g(y))+D_a g(z) = f(x)+D_a g(f(y)+D_a g(z)) \tag{8.7}$$

Example 8.1 *Let $f(x)=x^3$ and $g(x)=x^2$ over $\mathbb{F}_{3^2}$ be permutation and planar function, respectively. Then we simplify Equation* (11.3),

$$x^9+2x^3+2a^3y^3+ay^3+2az+2a^2z+a^6+a^3=0 \tag{8.8}$$

Case (i) *When $a=2$, we have the form of Equation* (11.4)

$$x^9+2x^3=0$$

The only possible values of $x=0,1,2$ are satisfied the above equation. It is also independent from y and z. Therefore total number of associative triples is $a(Q)=3\cdot 9\cdot 9=243$.

Case (ii) *When $a=1$, we have the form of Equation* (11.4)

$$x^9+2x^3+z+2=0$$

The only possible values of z among these $\{2\alpha, \alpha+2, 1\}$ (whenever $\alpha^2+2\alpha+2$ is irreducible polynomial) are satisfied the above equation for given $x \in \mathbb{F}_{3^2}$. It is also independent from y. For given $y \in \mathbb{F}_{3^2}$, the form of associative triples are

$$\begin{array}{ccc} (2\alpha+1,y,2\alpha) & (\alpha,y,\alpha+2) & (0,y,1) \\ (2\alpha,y,2\alpha) & (\alpha+1,y,\alpha+2) & (2,y,1) \\ (2\alpha+2,y,2\alpha) & (\alpha+2,y,\alpha+2) & (1,y,1) \end{array}$$

Therefore total number of associative triples is $a(Q)=9\cdot 9=81$.

In above example, we see that $a \in \mathbb{F}_q^*$ control the count of associative triples. Similarly we check for another values of $a \in \mathbb{F}_q^*$. We compute the associative triples for another values of $a \in \mathbb{F}_q^*$ by SageMath. And we get $a(Q) = N^2$, where $N = 3^2$.

In this way we take another combinations of f (permutation) and g (planar function) over $\mathbb{F}_q$ and find the number of solutions of Equation (11.3). Here these solutions are associated with number of associative triples of quasigroup (11.1).

8.4 CONCLUSION AND FUTURE SCOPE

Quasigroups with $a(Q) \leq n^2$ are cryptographically more important, where $|Q| = n$. In this paper, we show that there exist a combination of permutation and planar function over $\mathbb{F}_q$ which provide the quasigroups with $a(Q) = n^2$. For few cases, we count the associative triples by using SageMath. In future we can proceed our calculation with the other known permutations and planar functions over $\mathbb{F}_q$ which are cryptographically more sound and also develop a efficient method or technique for Equation (11.3). We can also represent quasigroups as vectorial Boolean functions, for more details we refer to [2, 3, 4]. Therefore we can find the relationship between other cryptographic characteristics, i.e., Walsh transform, nonlinearity and differential uniformity, etc., of obtained quasigroups with using permutation and planar function.

Funding
No funding was received for conducting this study.

Compliance with ethical standards
Conflict of interest The authors declare that they have no conflict of interest.

Authors' contributions
Tajender Kumar (Corresponding author) had the idea for the article, then I performed the literature search and drafted and critically revised the work.

REFERENCES

1. E. Biham and A. Shamir, Differential cryptanalysis of DES-like cryptosystems, *Journal of Cryptology*, **4(1)**, pp. 3-72, (1991).
2. A. Canteaut, Lecture Notes on Cryptographic Boolean Functions, *Inria, Paris, France*, version: March 10, 2016.
3. C. Carlet, Vectorial Boolean functions for cryptography, *Boolean Models and Methods in Mathematics, Computer Science, and Engineering*, **134**, 398-469, (2010).
4. C. Carlet and S. Mesnager, A note on Semi-Bent Boolean Functions, *IACR Cryptology ePrint Archive*, **486**, (2010).
5. R. S. Coulter and R. W. Matthews, Planar functions and planes of Lenz-Barlotti Class II, *Designs, Codes Cryptography*, **10**, 16-184, (1997).
6. R. S. Coulter and R. W. Matthews, Dembowski-Ostrom polynomials from Dickson polynomials, *Finite Fields and Their Applications*, textbf16(5), 369-379, (2010).
7. P. Dembowski and T. G. Ostrom, Planes of order n with collineation groups of order n^2, *Mathematische Zeitschrift*, **103**, 239-258, (1968).
8. A. Drapal, On Quasigroup Rich in Associative Triples, *Discrete Mathematics*, **44(3)**, 251-265, (1983).
9. A. Drapal and V. Valent, Few associative triples, isotopisms and groups, *Designs, Codes and Cryptography*, **86(3)**, 555-568, (2018).
10. O. Grošek and P. Horak, On Quasigroups with Few Associative Triples, *Designs, Codes and Cryptography*, **64**, 221-227, (2012).
11. J. Ježek and T. Kepka, Notes on the Number of Associative Triples, *Acta Universitatis Carolinae. Mathematica et Physica*, **31**, 15-19, (1990).
12. C. Koscielny, Generating Quasigroups for Cryptograhic Applications, *International Journal of Applied Mathematics and Computer Science*, **12**, 559-569, (2002).

13. A. Kotzig and C. Reischer, Associativity index of finite quasigroup, *Glasnik Mathematicki*, **18**, 243-253, (1983).
14. R. Lidl and H. Niederreiter, Finite Fields, Encyclopedia of Mathematics and Its Applications, *Addison - Wesley Publication*, (1983).
15. S. Markovski, Design of crypto primitives based on quasigroups, *Quasigroups and Related Systems*, **23**, 41-90, (2015).
16. A. Muratovic-Ribic and E. Pasalic. On derivatives of planar mappings and their connections to complete mappings, *Discrete Applied Mathematics*, **250**, 285-290, (2018).
17. D. A. Norton, A Note on Associativity, *Pacific Journal Of Mathematics*, **10**, 591-595, (1960).
18. A. Sade, Quasigroups automorphes par le groupe cylique, *Canadian Journal of Mathematics*, **9**, 321-335, (1957).
19. M. V. K. Satti, Quasigroup Based Cryptographic System, *A Thesis*, (2007).
20. V. Snasel, A. Abraham, J. Dvorsky, P. Krömer and J. Platos, Hash functions based on large quasigroups, *Computational Science-ICCS 2009*, 521-529, (2009).

9 Results on Arithmetic Statistically Convergence of Double Sequences on Intuitionistic Fuzzy Normed Spaces

Omer Kisi
Bartin University, Faculty of Sciences, Department of Mathematics, Kutlubey Campus, Bartin, Turkey

Erhan Guler
Bartin University, Faculty of Sciences, Department of Mathematics, Kutlubey Campus, Bartin, Turkey

CONTENTS

The study of summability theory and convergence of sequences in different spaces has become a significant topic in mathematical analysis. In the literature, researchers have presented some elegant works on intuitionistic fuzzy normed linear spaces. This work focuses on various types of convergence that are relevant to intuitionistic fuzzy normed spaces. The study combines arithmetic convergence with different forms of statistical convergence, which are then integrated into the structure of intuitionistic fuzzy normed spaces through the membership function. Initially, the concepts of arithmetic convergence and arithmetic statistical convergence are introduced in the intuitionistic fuzzy normed space. Then, the study examines important definitions that can be established with lacunary sequences and ideal structures, as well as the relationships between convergent sequences in this context. Additionally, new convergence definitions are established by evaluating lambda sequences together with arithmetic and statistical convergence, using the properties of the intuitionistic fuzzy normed space structure. Finally, the inclusion relation between the two sets is given with the help of the definition of graduated convergence.

9.1 INTRODUCTION

Using classical methods, explaining the numerous uncertainties that arise in our daily lives is not an easy task. To eliminate these ambiguities, fuzzy concepts were first investigated, followed by

DOI: 10.1201/9781032634142-9

intuitionistic fuzzy concepts, and finally neutrosophic concepts emerged. The development of fuzzy set theory, which employs the membership function to convert the elements of a set into the interval $[0,1]$, facilitated the evolution of intuitionistic fuzzy set theory by incorporating the non-membership function that also converts the elements of the set into the interval $[0,1]$. In this theory, the sum of non-membership and membership function values for all elements of the set always remains within the range of $[0,1]$.

Zadeh [31] initially introduced the theory of Fuzzy Sets (FSs), which has a significant impact on various scientific disciplines by enabling the utilization of imprecise and indistinct data. Despite the widespread implementation of FSs, they may not always adequately handle ambiguous membership degrees. Consequently, Atanassov [1] extended the theory of FSs by proposing the theory of Intuitionistic Fuzzy Sets (IFS). Additionally, Kramosil and Michalek [14] investigated Fuzzy Metric Spaces (FMS), which employ fuzzy and probabilistic metrics. Kaleva and Seikkala [9] analyzed FMSs as non-negative fuzzy numbers between two points. George and Veeramani [7] researched the properties of FMSs, proving substantial theorems and presenting crucial features. FMSs have practical applications in decision-making, fixed point theory, and medical imaging. Saadati and Park [22] worked on intuitionistic fuzzy topological spaces. Park [20] generalized FMSs by defining IF Metric Spaces (IFMS), following George and Veeramani's recommendation of utilizing t-norm and t-conorm. Lael and Nourouzi [15] investigated the concept of IF-Normed Spaces (IFNS) to establish a distinct topology from the topology generated by the F-norm μ, by eliminating the requirement $\mu + \nu \leq 1$ from Park's definition. Research conducted by [10, 12, 18, 19] has revealed several studies on the convergence of sequences in normed linear spaces with a fuzzy setting.

The concept of statistical convergence for a sequence of real numbers was initially introduced by Fast [6]. Mursaleen and Edely [17] extended this concept to double sequences.

The notion of λ-sequences, introduced by Mursaleen [16], has been combined with statistical convergence. Later, Savaş [23] proposed λ-statistical convergence of fuzzy numbers. Belen and Yildirim investigated a generalization of statistical convergence, (λ,μ)-statistical convergence, (V,λ,μ) summability, and $(C,1,1)$ summability for double sequences using ideals in their work [2]. Additionally, Esi and Hazarika [5], Hazarika and Savaş [8], and Savaş and Patterson [24] explored and examined double λ-statistical convergence.

Ideal convergence was initially examined by Kostyrko et al. [13], and since then, it has been extensively studied from various angles in fuzzy, intuitionistic fuzzy, and neutrosophic normed spaces. Das et al. [3] reconsidered the concept of $\mathscr{I}$-convergence of double sequences and established some properties of this type of convergence. $\mathscr{I}$-convergence has gained significant attention in summability theory following the research of [4].

On the other hand, Ruckle [21] proposed the concept of arithmetic convergence, which has since been worked by various authors in relation to arithmetic convergence, statistical convergence, metric spaces, and invariance. This concept has been explored in works by Kişi [11] and Yaying and Hazarika [26, 27, 28, 29, 30].

After a thorough analysis of existing literature, this study addresses a significant research gap by investigating the transferability of the concept of arithmetic convergence to spaces with intuitionistic norms. By incorporating various related concepts such as statistical convergence, lambda, ideal, and lacunary convergence, the study aims to establish a comprehensive framework within the realm of IFNS. The proposed framework not only introduces the definitions of convergence for these concepts but also examines their inclusion status, thereby enhancing our understanding of their interplay in IFNS.

9.2 PRELIMINARIES

In this section, we give significant existing conceptions and results which are crucial for our findings.

Definition 9.1 *A sequence (w_u) is called to be arithmetic statistically convergent to $w_{\langle u,s\rangle}$ if, for all $\rho > 0$, there is an integer s so that*

$$\lim_{t\to\infty}\frac{1}{t}\left|\left\{u\le t : \left|w_u - w_{\langle u,s\rangle}\right|\ge\rho\right\}\right| = 0.$$

The notation $\lim w_u = w_{\langle u,s\rangle}(ASt)$ is used to represent the arithmetic statistical convergence of (w_u) to $w_{\langle u,s\rangle}$. The symbol $\langle u,s\rangle$ indicates the greatest common divisor of u and s, which is defined as the largest number that divides both u and s.

Definition 9.2 *(w_u) is named to be lacunary arithmetic statistically convergent to $w_{\langle u,s\rangle}$ if for a lacunary sequence $\theta = (k_t)$ and for all $\rho > 0$, there is an integer s so that*

$$\lim_{t\to\infty}\frac{1}{t}\left|\left\{u\in\mathscr{I}_t : \left|w_u - w_{\langle u,s\rangle}\right|\ge\rho\right\}\right| = 0.$$

Then, it is demonstrated with $\lim w_u = w_{\langle u,s\rangle}(ASt_\theta)$. Here, (θ) is an increasing integer sequence, $k_0 = 0, \mathfrak{h}_t = k_t - k_{t-1}\to\infty$ as $t\to\infty$ and $\mathscr{I}_t = (k_{t-1}, k_t]$.

The concept of continuous *t-norm* and *t-conorm* was put forward by Schweizer and Sklar [25]. Saadati and Park [22] have recently proposed the notion of IFNS by utilizing the continuous *t-norm* and *t-conorm*. This novel concept can be described as follows:

Definition 9.3 *Let (Y,ψ,ϕ) denote a triplicate, where Y represents a real vector space, $\diamond$ be continuous t-norm, $*$ be continuous t-conorm, and ψ and ϕ are F-sets on $X\times F$ that satisfy the following statements for all $u,v\in Y$ and $j,k\in\mathbb{R}$:*
(a) $\psi(u,j)+\phi(u,j)\le 1$,
(b) $\psi(u,j) > 0$,
(c) $\psi(u,j) = 1$ *iff* $u = 0$,
(d) $\psi(qu,j) = \psi\left(u,\frac{j}{|q|}\right)$ *for each* $q\ne 0$,
(e) $\psi(u,j)\diamond\psi(v,k)\le\psi(u+v,j+k)$,
(f) $\psi(u,.) : (0,\infty)\to[0,1]$ *is continuous,*
(g) $\lim_{j\to\infty}\psi(u,j) = 1$ *and* $\lim_{j\to 0}\psi(u,j) = 0$,
(h) $\phi(u,j) < 1$;
(i) $\phi(u,j) = 0$ *iff* $u = 0$,
(j) $\phi(qu,j) = \phi\left(u,\frac{j}{|q|}\right)$ *for each* $q\ne 0$,
(k) $\phi(u,j)*\phi(v,k)\ge\phi(u+v,j+k)$,
(l) $\phi(u,.) : (0,\infty)\to[0,1]$ *is continuous,*
(m) $\lim_{j\to\infty}\phi(u,j) = 0$ *and* $\lim_{j\to 0}\phi(u,j) = 1$.
In this context, (ψ,ϕ) is referred to as an IF-norm on Y. Furthermore, (Y,ψ) is termed an F-normed space, with ϕ serving as an additional component.

A double sequence $\theta_2 = \{(j_m,k_n)\}$ is referred to as a double lacunary sequence if there exist two increasing sequences of integers (j_m) and (k_n) satisfying the following conditions:
1) $j_0 = 0$ and the difference $h_m = j_m - j_{m-1}$ tends to infinity as u approaches infinity.
2) $k_0 = 0$ and the difference $\overline{h}_n = k_n - k_{n-1}$ tends to infinity as n approaches infinity.
In the context of this discussion, the following notations will be used:

$$j_{mn} := j_m k_n,\ \mathfrak{h}_{m,n} := h_m\overline{h}_n \text{ and}$$

θ_2 is determined by the set

$$\mathscr{I}_{m,n} := \{(j,k) : j_{m-1} < j\le j_m,\ k_{n-1} < k\le k_n\},$$

$$q_m := \frac{j_m}{j_{m-1}},\ \overline{q}_n := \frac{k_n}{k_{n-1}} \text{ and } q_{mn} := q_m\overline{q}_n.$$

A double sequence $(w_{u,v})$ of numbers is defined to be $\mathcal{I}_2$-lacunary statistical convergent or $S_{\theta_2}(\mathcal{I}_2)$-convergent to w_0, provided that for all $\rho,\sigma>0$,

$$\left\{(m,n)\in\mathbb{N}\times\mathbb{N}:\frac{1}{\mathfrak{h}_{m,n}}|\{(u,v)\in\mathcal{J}_{m,n}:|w_{u,v}-w_0|\geq\rho\}|\geq\sigma\right\}\in\mathcal{I}_2.$$

In this situation, we denote the convergence of $(w_{u,v})$ to w_0 under $S_{\theta_2}(\mathcal{I}_2)$ as $w_{u,v}\to w_0\left(S_{\theta_2}(\mathcal{I}_2)\right)$.

9.3 MAIN RESULTS

Definition 9.4 *Assume $(Y,\psi,\phi,\diamond,*)$ be an IFNS. $(w_{u,v})$ is said to be a arithmetic convergent to $w_{\langle u,s\rangle,\langle v,t\rangle}$ on $(Y,\psi,\phi,\diamond,*)$ provided that for each $\rho\in(0,1)$, $q>0$ and there are integers s,t, and $k^*,l^*\in\mathbb{N}$, such that for all $u\geq k^*,v\geq l^*$,*

$$\psi\left(w_{u,v}-w_{\langle u,s\rangle,\langle v,t\rangle},q\right)>1-\rho \text{ and } \phi\left(w_{u,v}-w_{\langle u,s\rangle,\langle v,t\rangle},q\right)<\rho. \tag{9.1}$$

Symbolically, it is denoted as $(\psi,\phi)-\lim w_{u,v}=w_{\langle u,s\rangle,\langle v,t\rangle}$. Here, (9.1) implies

$$\lim_{u,v\to\infty}\psi\left(w_{u,v}-w_{\langle u,s\rangle,\langle v,t\rangle},q\right)=1,\ \lim_{u,v\to\infty}\phi\left(w_{u,v}-w_{\langle u,s\rangle,\langle v,t\rangle},q\right)=0. \tag{9.2}$$

Lemma 9.1 *Assume $(Y,\psi,\phi,\diamond,*)$ be an IFNS. The subsequent features are valid.*
1) $(w_{u,v})$ is arithmetic convergent on $(Y,\psi,\phi,\diamond,*)$ then it is unique.
2) Assume $\lim_{u,v\to\infty}w_{u,v}=w_{\langle u,s\rangle,\langle v,t\rangle}$ and $\lim_{u,v\to\infty}\breve{w}_{u,v}=\breve{w}_{\langle u,s\rangle,\langle v,t\rangle}$. Then

$$\lim_{u,v\to\infty}w_{u,v}+\breve{w}_{u,v}=w_{\langle u,s\rangle,\langle v,t\rangle}+\breve{w}_{\langle u,s\rangle,\langle v,t\rangle}$$

on $(Y,\psi,\phi,\diamond,*)$.
3) Assume $\lim_{u,v\to\infty}w_{u,v}=w_{\langle u,s\rangle,\langle v,t\rangle}$ and $c\neq0$, then $\lim_{u,v\to\infty}cw_{u,v}=cw_{\langle u,s\rangle,\langle v,t\rangle}$.

Proof *The proof is easily obtained from the (9.1) or (9.2).*

The definition of arithmetic convergence on $(Y,\psi,\phi,\diamond,*)$ can be modified to include the concept of statistical convergence, as shown below.

Definition 9.5 *A sequence $(w_{u,v})$ is said to be a arithmetic statistically convergent to $w_{\langle u,s\rangle,\langle v,t\rangle}$ on $(Y,\psi,\phi,\diamond,*)$ provided that for all $\rho\in(0,1)$, $q>0$ and there are integers s,t, such that*

$$\lim_{m,n\to\infty}\frac{1}{mn}\left|\left\{u\leq m,v\leq n:\psi\left(w_{u,v}-w_{\langle u,s\rangle,\langle v,t\rangle},q\right)\leq1-\rho \text{ or } \phi\left(w_{u,v}-w_{\langle u,s\rangle,\langle v,t\rangle},q\right)\geq\rho\right\}\right|=0. \tag{9.3}$$

Symbolically, it can be represented as $(\psi,\phi)-\lim w_{u,v}=w_{\langle u,s\rangle,\langle v,t\rangle}(ASt)$.

The definition given by equation (9.3) can be used in the following equivalent statement:

Lemma 9.2 *Suppose $(Y,\psi,\phi,\diamond,*)$ be an IFNS. The following statements are equivalent:*
1)

$$(\psi,\phi)-\lim w_{u,v}=w_{\langle u,s\rangle,\langle v,t\rangle}(ASt).$$

2)

$$\lim_{m,n\to\infty}\tfrac{1}{mn}\left|\left\{u\leq m,v\leq n:\psi\left(w_{u,v}-w_{\langle u,s\rangle,\langle v,t\rangle},q\right)\leq1-\rho\right\}\right|$$

$$=\lim_{m,n\to\infty}\tfrac{1}{mn}\left|\left\{u\leq m,v\leq n:\phi\left(w_{u,v}-w_{\langle u,s\rangle,\langle v,t\rangle},q\right)\geq\rho\right\}\right|=0.$$

3)

$$\lim_{u,v\to\infty}\psi\left(w_{u,v}-w_{\langle u,s\rangle,\langle v,t\rangle},q\right)=1(ASt),\ \lim_{u,v\to\infty}\phi\left(w_{u,v}-w_{\langle u,s\rangle,\langle v,t\rangle},q\right)=0(ASt).$$

Proof *The proof can be readily obtained from the definition. The following lemma presents the relationship between the two definitions mentioned above.*

Lemma 9.3 *Let* $(Y,\psi,\phi,\diamond,*)$ *be an IFNS and* $(\psi,\phi)-\lim w_{u,v}=w_{\langle u,s\rangle,\langle v,t\rangle}$. *In this case,* $(\psi,\phi)-\lim w_{u,v}=w_{\langle u,s\rangle,\langle v,t\rangle}(ASt)$.

Proof *Let* $(\psi,\phi)-\lim w_{u,v}=w_{\langle u,s\rangle,\langle v,t\rangle}$, *so for each* $\rho\in(0,1), q>0$ *and there are integers* s,t, *and* $k^*,l^*\in\mathbb{N}$, *so that for all* $u\geq k^*, v\geq l^*$,

$$\psi\left(w_{u,v}-w_{\langle u,s\rangle,\langle v,t\rangle},q\right)>1-\rho \text{ and } \phi\left(w_{u,v}-w_{\langle u,s\rangle,\langle v,t\rangle},q\right)<\rho.$$

So, the density of

$$\left\{u\leq m, v\leq n:\psi\left(w_{u,v}-w_{\langle u,s\rangle,\langle v,t\rangle},q\right)\leq 1-\rho \text{ or } \phi\left(w_{u,v}-w_{\langle u,s\rangle,\langle v,t\rangle},q\right)\geq\rho\right\}$$

is zero. Hence

$$\lim_{m,n\to\infty}\frac{1}{mn}\mid\left\{u\leq m, v\leq n:\psi\left(w_{u,v}-w_{\langle u,s\rangle,\langle v,t\rangle},q\right)\leq 1-\rho \text{ or } \phi\left(w_{u,v}-w_{\langle u,s\rangle,\langle v,t\rangle},q\right)\geq\rho\mid=0.\right.$$

Namely, $(\psi,\phi)-\lim_{u,v}=w_{\langle u,s\rangle,\langle v,t\rangle}(ASt)$.

The following hybrid definitions integrate the concepts of arithmetic convergence and lacunary convergence.

Definition 9.6 *Assume* $(Y,\psi,\phi,\diamond,*)$ *be an IFNS,* $\theta_2=\{(j_m,k_n)\}$ *be lacunary sequence.* $(w_{u,v})$ *is named to be arithmetic strong lacunary convergent to* $w_{\langle u,s\rangle,\langle v,t\rangle}$ *on* $(Y,\psi,\phi,\diamond,*)$ *provided that for each* $\rho\in(0,1), q>0$, *there are integers* s,t *and* $k^*,l^*\in\mathbb{N}$, *such that for all* $u\geq k^*, v\geq l^*$

$$\frac{1}{\mathfrak{h}_{mn}}\sum_{u,v\in\mathscr{I}_{m,n}}\psi\left(w_{u,v}-w_{\langle u,s\rangle,\langle v,t\rangle},q\right)>1-\rho \text{ and } \frac{1}{\mathfrak{h}_{mn}}\sum_{u,v\in\mathscr{I}_{m,n}}\phi\left(w_{u,v}-w_{\langle u,s\rangle,\langle v,t\rangle},q\right)<\rho.$$

Symbolically, it can be represented as $(\psi,\phi)-\lim w_{u,v}=w_{\langle u,s\rangle,\langle v,t\rangle}\left(A\theta_2^S\right)$.

Theorem 9.1 *If* $(\psi,\phi)-\lim_{u,v}=w_{\langle u,s\rangle,\langle v,t\rangle}\left(A\theta_2^S\right)$, *in this case, this limit is unique.*

Proof *Let's assume that*

$$(\psi,\phi)-\lim_{u,v}=w_{\langle u,s\rangle,\langle v,t\rangle}\left(A\theta_2^S\right),$$

and

$$(\psi,\phi)-\lim_{u,v}=\breve{w}_{\langle u,s\rangle,\langle v,t\rangle}\left(A\theta_2^S\right)$$

For given $\rho\in(0,1)$, *selecting* $m\in(0,1)$ *so that* $(1-m)\diamond(1-m)>1-\rho$ *and* $m*m<\rho$. *For all* $q>0$, *there are integers* s,t, *and* $k^*,l^*\in\mathbb{N}$, *such that for each* $u\geq k^*, v\geq l^*$,

$$\frac{1}{\mathfrak{h}_{mn}}\sum_{u,v\in\mathscr{I}_{m,n}}\psi\left(w_{u,v}-w_{\langle u,s\rangle,\langle v,t\rangle},q\right)>1-\rho \text{ and } \frac{1}{\mathfrak{h}_{mn}}\sum_{u,v\in\mathscr{I}_{m,n}}\phi\left(w_{u,v}-w_{\langle u,s\rangle,\langle v,t\rangle},q\right)<\rho.$$

Also, there is $k^{**},l^{**}\in\mathbb{N}$ *such that, for each* $u\geq k^{**}, v\geq l^{**}$

$$\frac{1}{\mathfrak{h}_{mn}}\sum_{u,v\in\mathscr{I}_{m,n}}\psi\left(w_{u,v}-\breve{w}_{\langle u,s\rangle,\langle v,t\rangle},q\right)>1-\rho \text{ and } \frac{1}{\mathfrak{h}_{mn}}\sum_{u,v\in\mathscr{I}_{m,n}}\phi\left(w_{u,v}-\breve{w}_{\langle u,s\rangle,\langle v,t\rangle},q\right)<\rho.$$

Consider $\widehat{k} = \max\{k^*, k^{**}\}$, $\widehat{l} = \max\{l^*, l^{**}\}$. *So, for* $u \geq \widehat{k}, v \geq \widehat{l}$ *taking a* $(y,z) \in \mathbb{N} \times \mathbb{N}$ *such that*

$$\psi\left(w_{y,z} - w_{\langle u,s\rangle,\langle v,t\rangle}, \frac{q}{2}\right) > \frac{1}{\mathfrak{h}_{mn}} \sum_{u,v \in \mathscr{I}_{m,n}} \psi\left(w_{u,v} - w_{\langle u,s\rangle,\langle v,t\rangle}, q\right) > 1 - m$$

and

$$\psi\left(w_{y,z} - \breve{w}_{\langle u,s\rangle,\langle v,t\rangle}, \frac{q}{2}\right) > \frac{1}{\mathfrak{h}_{mn}} \sum_{u,v \in \mathscr{I}_{m,n}} \psi\left(w_{u,v} - \breve{w}_{\langle u,s\rangle,\langle v,t\rangle}, q\right) > 1 - m.$$

So,

$$\begin{aligned} \psi\left(w_{\langle u,s\rangle,\langle v,t\rangle} - \breve{w}_{\langle u,s\rangle,\langle v,t\rangle}, q\right) &> \psi\left(w_{y,z} - w_{\langle u,s\rangle,\langle v,t\rangle}, \tfrac{q}{2}\right) \diamond \psi\left(w_{y,z} - \breve{w}_{\langle u,s\rangle,\langle v,t\rangle}, \tfrac{q}{2}\right) \\ &> (1-m) \diamond (1-m) > 1 - \rho. \end{aligned}$$

Using $\rho > 0$ *is arbitrary. For all* $q > 0$, *if we write* $\lim_{n\to\infty} \psi\left(w_{u,v} - \breve{w}_{\langle u,s\rangle,\langle v,t\rangle}, q\right) = 1$ *and do the same for another, then we have* $w_{\langle u,s\rangle,\langle v,t\rangle} = \breve{w}_{\langle u,s\rangle,\langle v,t\rangle}$.

Definition 9.7 *Let* $(Y, \psi, \phi, \diamond, *)$ *be an IFNS,* $\theta_2 = \{(j_m, k_n)\}$ *be a lacunary sequence.* $(w_{u,v})$ *is called to be arithmetic lacunary statistically convergent to* $w_{\langle u,s\rangle,\langle v,t\rangle}$ *on* $(Y, \psi, \phi, \diamond, *)$ *provided that for each* $\rho \in (0,1), q > 0$ *and there are integers* s, t, *so that*

$$\lim_{m,n\to\infty} \frac{1}{\mathfrak{h}_{mn}} \left|\left\{(u,v) \in \mathscr{I}_{m,n} : \psi\left(w_{u,v} - w_{\langle u,s\rangle,\langle v,t\rangle}, q\right) \leq 1 - \rho \text{ or } \phi\left(w_{u,v} - w_{\langle u,s\rangle,\langle v,t\rangle}, q\right) \geq \rho\right\}\right| = 0. \tag{9.4}$$

Symbolically, it is denoted as $(\psi, \phi) - \lim w_{u,v} = w_{\langle u,s\rangle,\langle v,t\rangle}\left(ASt_{\theta_2}\right)$.

The lemma below provides alternative expressions that can be used in place of (9.4).

Lemma 9.4 *Let* $(Y, \psi, \phi, \diamond, *)$ *be an IFNS,* $\theta_2 = \{(j_m, k_n)\}$ *be a lacunary sequence. Then, the subsequent statements are equivalent.*

1)

$$(\psi, \phi) - \lim w_{u,v} = w_{\langle u,s\rangle,\langle v,t\rangle}\left(ASt_\theta\right).$$

2)

$$\begin{aligned} &\lim_{m,n\to\infty} \tfrac{1}{\mathfrak{h}_{mn}} \left|\left\{(u,v) \in \mathscr{I}_{m,n} : \psi\left(w_{u,v} - w_{\langle u,s\rangle,\langle v,t\rangle}, q\right) \leq 1 - \rho\right\}\right| \\ &= \lim_{m,n\to\infty} \tfrac{1}{\mathfrak{h}_{mn}} \left|\left\{(u,v) \in \mathscr{I}_{m,n} : \phi\left(w_{u,v} - w_{\langle u,s\rangle,\langle v,t\rangle}, q\right) \geq \rho\right\}\right| = 0. \end{aligned}$$

3)

$$\begin{aligned} &\lim_{m,n\to\infty} \tfrac{1}{\mathfrak{h}_{mn}} \left|\left\{(u,v) \in \mathscr{I}_{m,n} : \psi\left(w_{u,v} - w_{\langle u,s\rangle,\langle v,t\rangle}, , q\right) > 1 - \rho\right\}\right| \\ &= \lim_{m,n\to\infty} \tfrac{1}{\mathfrak{h}_{mn}} \left|\left\{(u,v) \in \mathscr{I}_{m,n} : \phi\left(w_{u,v} - w_{\langle u,s\rangle,\langle v,t\rangle}, q\right) < \rho\right\}\right| = 1. \end{aligned}$$

4)

$$(\psi, \phi) - \lim \psi\left(w_{u,v} - w_{\langle u,s\rangle,\langle v,t\rangle}, q\right) = 1\left(ASt_\theta\right),$$

$$(\psi, \phi) - \lim \phi\left(w_{u,v} - w_{\langle u,s\rangle,\langle v,t\rangle}, q\right) = 0\left(ASt_\theta\right).$$

Definition 9.8 *A sequence* $(w_{u,v})$ *is called to be a arithmetic* λ*-statistically convergent to* $w_{\langle u,s\rangle,\langle v,t\rangle}$ *on* $(Y, \psi, \phi, \diamond, *)$ *provided that for all* $\rho \in (0,1), q > 0$ *and there are integers* s, t, *such that*

$$\lim_{m,n\to\infty} \frac{1}{\lambda_m \mu_n} \left|\left\{(u,v) \in J_m \times I_n : \psi\left(w_{u,v} - w_{\langle u,s\rangle,\langle v,t\rangle}, q\right) \leq 1 - \rho \text{ or } \phi\left(w_{u,v} - w_{\langle u,s\rangle,\langle v,t\rangle}, q\right) \geq \rho\right\}\right| = 0. \tag{9.5}$$

Symbolically, it can be represented as $(\psi, \phi) - \lim w_{u,v} = w_{\langle u,s\rangle,\langle v,t\rangle}\left(ASt_{\lambda,\mu}\right)$.
Here, $\lambda = (\lambda_m) \in \Delta$, $\mu = (\mu_n) \in \Delta$, $J_m = [m - \lambda_m + 1, m]$ *and* $I_n = [n - \mu_n + 1, n]$.

Definition 9.9 *A sequence $(w_{u,v})$ is defined as arithmetic $\mathcal{I}_2$-statistically convergent to $w_{\langle u,s\rangle,\langle v,t\rangle}$ on $(Y,\psi,\phi,\diamond,*)$ provided that for all $\rho\in(0,1),\sigma>0,q>0$ and there are integers s,t, such that*

$$\Big\{(m,n)\in\mathbb{N}\times\mathbb{N}:\tfrac{1}{mn}\big|\{u\leq m,v\leq n:\psi\left(w_{u,v}-w_{\langle u,s\rangle,\langle v,t\rangle},q\right)\leq 1-\rho \\ \text{or } \phi\left(w_{u,v}-w_{\langle u,s\rangle,\langle v,t\rangle},q\right)\geq\rho\}\big|\geq\sigma\Big\}\in\mathcal{I}_2. \tag{9.6}$$

Then, it is demonstrated as $(\psi,\phi)-\lim w_{u,v}=w_{\langle u,s\rangle,\langle v,t\rangle}\left(ASt_{\mathcal{I}_2}\right)$.

Definition 9.10 *A sequence $(w_{u,v})$ is named to be arithmetic $\lambda\mathcal{I}_2$-statistically convergent to $w_{\langle u,s\rangle,\langle v,t\rangle}$ on $(Y,\psi,\phi,\Delta,*)$ provided that for all $\rho\in(0,1),\sigma>0,q>0$ and there are integers s,t, such that*

$$\Big\{(m,n)\in\mathbb{N}\times\mathbb{N}:\tfrac{1}{\lambda_m\mu_n}\big|\{(u,v)\in J_m\times I_n:\psi\left(w_{u,v}-w_{\langle u,s\rangle,\langle v,t\rangle},q\right)\leq 1-\rho \\ \text{or } \phi\left(w_{u,v}-w_{\langle u,s\rangle,\langle v,t\rangle},q\right)\geq\rho\}\big|\geq\sigma\Big\}\in\mathcal{I}_2. \tag{9.7}$$

Symbolically, it can be represented as $(\psi,\phi)-\lim w_{u,v}=w_{\langle u,s\rangle,\langle v,t\rangle}\left(ASt_{\lambda\mathcal{I}}\right)$.

The definition of arithmetic $\lambda\mathcal{I}_2$-statistically summability can be provided by utilizing the Valee-Pousin sum, also known as $t_{mn}(w)$ in literature, which is given by the formula:

$$t_{uv}(w)=\frac{1}{\lambda_u\mu_v}\sum_{(m,n)\in J_u\times I_v}w_{m,n}$$

Definition 9.11 *A sequence $(w_{u,v})$ is said to be a arithmetic $\lambda\mathcal{I}_2$-statistically summable to $w_{\langle u,s\rangle,\langle v,t\rangle}$ on $(Y,\psi,\phi,\diamond,*)$ provided that for all $\rho\in(0,1),\sigma>0,q>0$ and there are integers s,t, such that*

$$\Big\{(m,n)\in\mathbb{N}\times\mathbb{N}:\tfrac{1}{\lambda_m\mu_n}\big|\{(u,v)\in J_m\times I_n:\psi\left(t_{uv}(w)-w_{\langle u,s\rangle,\langle v,t\rangle},q\right)\leq 1-\rho \\ \text{or } \phi\left(t_{uv}(w)-w_{\langle u,s\rangle,\langle v,t\rangle},q\right)\geq\rho\}\big|\geq\sigma\Big\}\in\mathcal{I}_2. \tag{9.8}$$

Symbolically, it can be represented as $(\psi,\phi)-[\lambda\mathcal{I}_2]-\lim w_{u,v}=w_{\langle u,s\rangle,\langle v,t\rangle}\left(ASt_{\lambda\mathcal{I}_2}\right)$.

We can state the following theorem to illustrate the connection between arithmetic $\lambda\mathcal{I}_2$ statistically summable and arithmetic $\lambda\mathcal{I}_2$-statistically convergent.

Theorem 9.2 *If $(w_{u,v})$ is a arithmetic $\lambda\mathcal{I}_2$-statistically summable to $w_{\langle u,s\rangle,\langle v,t\rangle}$ then $(w_{u,v})$ is arithmetic $\lambda\mathcal{I}_2$-statistically convergent to $w_{\langle u,s\rangle,\langle v,t\rangle}$.*

Proof *Let $(w_{u,v})$ is a arithmetic $\lambda\mathcal{I}_2$-statistically summable to $w_{\langle u,s\rangle,\langle v,t\rangle}$ on $(Y,\psi,\phi,\diamond,*)$. Then*

$$\sum_{(u,v)\in J_m\times I_n}\left(\psi\left(w_{u,v}-w_{\langle u,s\rangle,\langle v,t\rangle},q\right)\text{ or }\phi\left(w_{u,v}-w_{\langle u,s\rangle,\langle v,t\rangle},q\right)\right)$$

$$\geq\sum_{\substack{(u,v)\in J_m\times I_n,\psi\left(w_{u,v}-w_{\langle u,s\rangle,\langle ,t\rangle},q\right)\leq 1-\rho\\ \phi\left(w_{u,v}-w_{\langle u,s\rangle,\langle v,t\rangle},q\right)\geq\rho}}\left(\psi\left(w_{u,v}-w_{\langle u,s\rangle,\langle v,t\rangle},q\right)\text{ or }\phi\left(w_{u,v}-w_{\langle u,s\rangle,\langle v,t\rangle},q\right)\right)$$

$$\geq\rho.\left|\{(u,v)\in J_m\times I_n:\psi\left(w_{u,v}-w_{\langle u,s\rangle,\langle v,t\rangle},q\right)\leq 1-\rho\text{ or }\phi\left(w_{u,v}-w_{\langle u,s\rangle,\langle v,t\rangle},q\right)\geq\rho\}\right|.$$

So,

$$\frac{1}{\lambda_m\mu_n}\left|\left\{(u,v)\in J_m\times I_n:\psi\left(w_{u,v}-w_{\langle u,s\rangle,\langle v,t\rangle},q\right)\leq 1-\rho \text{ or } \phi\left(w_{u,v}-w_{\langle u,s\rangle,\langle v,t\rangle},q\right)\geq\rho\right\}\right|\geq\sigma.$$

Hence, we obtain

$$\frac{1}{\lambda_m\mu_n}\sum_{(u,v)\in J_m\times I_n}\psi\left(w_{u,v}-w_{\langle u,s\rangle,\langle v,t\rangle},q\right)\leq(1-\rho)\sigma \text{ or } \phi\left(w_{u,v}-w_{\langle u,s\rangle,\langle v,t\rangle},q\right)\geq\rho\sigma.$$

Therefore, we write

$$\Big\{(m,n)\in\mathbb{N}\times\mathbb{N}:\tfrac{1}{\lambda_m\mu_n}\left|\left\{(u,v)\in J_m\times I_n:\psi\left(w_{u,v}-w_{\langle u,s\rangle,\langle v,t\rangle},q\right)\leq 1-\rho\right.\right.$$

$$\left.\left.\text{or } \phi\left(w_{u,v}-w_{\langle u,s\rangle,\langle v,t\rangle},q\right)\geq\rho\right\}\right|\geq\sigma\Big\}$$

$$\subset\Big\{(m,n)\in\mathbb{N}\times\mathbb{N}:\tfrac{1}{\lambda_m\mu_n}\textstyle\sum_{(u,v)\in J_m\times I_n}\psi\left(w_{u,v}-w_{\langle u,s\rangle,\langle v,t\rangle},q\right)\leq(1-\rho)\sigma$$

$$\text{or } \phi\left(w_{u,v}-w_{\langle u,s\rangle,\langle v,t\rangle},q\right)\geq\rho\sigma\Big\}.$$

As a result, when $(w_{u,v})$ *is a arithmetic* $\lambda\mathscr{I}_2$*-statistically summable to* $w_{\langle u,s\rangle,\langle v,t\rangle}$ *then* $(w_{u,v})$ *is arithmetic* $\lambda\mathscr{I}_2$*-statistically convergent to* $w_{\langle u,s\rangle,\langle v,t\rangle}$.

The theorem below demonstrates the inclusion connection between arithmetic $\mathscr{I}_2$-statistically convergent and arithmetic $\lambda\mathscr{I}_2$-statistically convergent.

Arithmetic $\mathscr{I}_2$-statistically convergent sequences and arithmetic $\lambda\mathscr{I}_2$-statistically convergent sequences spaces are demonstrated with $S\left(ASt_{\mathscr{I}_2}\right)$ and $S\left(ASt_{\lambda\mathscr{I}_2}\right)$, respectively.

Theorem 9.3 *If* $\lim_{m,n\to\infty}\frac{\lambda_m\mu_n}{mn}>0$, *then* $S\left(ASt_{\mathscr{I}_2}\right)\subset S\left(ASt_{\lambda\mathscr{I}_2}\right)$

Proof *Given* $\rho>0$,

$$\tfrac{1}{mn}\left|\left\{u\leq m,v\leq n:\psi\left(w_{u,v}-w_{\langle u,s\rangle,\langle v,t\rangle},q\right)\leq 1-\rho \text{ or } \phi\left(w_{u,v}-w_{\langle u,s\rangle,\langle v,t\rangle},q\right)\geq\rho\right\}\right|$$

$$\geq\tfrac{1}{mn}\left|\left\{(u,v)\in J_m\times I_n:\psi\left(w_{u,v}-w_{\langle u,s\rangle,\langle v,t\rangle},q\right)\leq 1-\rho \text{ or } \phi\left(w_{u,v}-w_{\langle u,s\rangle,\langle v,t\rangle},q\right)\geq\rho\right\}\right|$$

$$=\tfrac{\lambda_m\mu_n}{mn}\tfrac{1}{\lambda_m\mu_n}\left|\left\{(u,v)\in J_m\times I_n:\psi\left(w_{u,v}-w_{\langle u,s\rangle,\langle v,t\rangle},q\right)\leq 1-\rho \text{ or } \phi\left(w_{u,v}-w_{\langle u,s\rangle,\langle v,t\rangle},q\right)\geq\rho\right\}\right|.$$

Here, utilizing $\lim_{m,n\to\infty}\frac{\lambda_m\mu_n}{mn}=d>0$, *then*

$$\left\{(m,n)\in\mathbb{N}\times\mathbb{N}:\frac{\lambda_m\mu_n}{mn}<\frac{d}{2}\right\}$$

is finite. So, for $\sigma>0$

$$\Big\{(m,n)\in\mathbb{N}\times\mathbb{N}:\tfrac{1}{\lambda_m\mu_n}\left|\left\{(u,v)\in J_m\times I_n:\psi\left(w_{u,v}-w_{\langle u,s\rangle,\langle v,t\rangle},q\right)\leq 1-\rho\right.\right.$$

$$\left.\left.\text{or } \phi\left(w_{u,v}-w_{\langle u,s\rangle,\langle v,t\rangle},q\right)\geq\rho\right\}\right|\geq\sigma\Big\}$$

$$\subset\Big\{(m,n)\in\mathbb{N}\times\mathbb{N}:\tfrac{1}{mn}\left|\left\{u\leq m,v\leq n:\psi\left(w_{u,v}-w_{\langle u,s\rangle,\langle v,t\rangle},q\right)\leq 1-\rho\right.\right.$$

$$\left.\left.\text{or } \phi\left(w_{u,v}-w_{\langle u,s\rangle,\langle v,t\rangle},q\right)\geq\rho\right\}\right|\geq\sigma\Big\}$$

$$\cup\left\{(m,n)\in\mathbb{N}\times\mathbb{N}:\tfrac{\lambda_m\mu_n}{mn}<\tfrac{d}{2}\right\}.$$

The reason the proof is complete is that the set on the right-hand side belongs to $\mathscr{I}_2$.

Theorem 9.4 *If* $\lim_{m\to\infty}\frac{\lambda_m\mu_n}{mn}=1$, *then* $S\left(ASt_{\lambda\mathscr{I}_2}\right)\subset S\left(ASt_{\mathscr{I}_2}\right)$.

Proof *Given for* $\sigma>0$, *let* $\lim_{m\to\infty}\frac{\lambda_m\mu_n}{mn}=1$. *Then, for all* $m\geq\check{m},n\geq\check{n}$, *there are* $\check{m},\check{n}\in\mathbb{N}$ *such that* $\left|\frac{\lambda_m\mu_n}{mn}-1\right|<\sigma$. *For all* $\rho>0$ *and* $m\geq\check{m},n\geq\check{n}$,

$$\frac{1}{mn}\left|\left\{u\leq m,v\leq n:\psi\left(w_{u,v}-w_{\langle u,s\rangle,\langle v,t\rangle},q\right)\leq 1-\rho\ or\ \phi\left(w_{u,v}-w_{\langle u,s\rangle,\langle v,t\rangle},q\right)\geq\rho\right\}\right|$$

$$=\frac{1}{mn}\left|\left\{u\leq m-\lambda_m,v\leq n-\mu_n:\psi\left(w_{u,v}-w_{\langle u,s\rangle,\langle v,t\rangle},q\right)\leq 1-\rho\ or\ \phi\left(w_{u,v}-w_{\langle u,s\rangle,\langle v,t\rangle},q\right)\geq\rho\right\}\right|$$

$$+\frac{1}{mn}\left|\left\{(u,v)\in J_m\times I_n:\psi\left(w_{u,v}-w_{\langle u,s\rangle,\langle v,t\rangle},q\right)\leq 1-\rho\ or\ \phi\left(w_{u,v}-w_{\langle u,s\rangle,\langle v,t\rangle},q\right)\geq\rho\right\}\right|$$

$$\leq\frac{mn-\lambda_m\mu_n}{mn}+\frac{1}{mn}\left|\left\{(u,v)\in J_m\times I_n:\psi\left(w_{u,v}-w_{\langle u,s\rangle,\langle v,t\rangle},q\right)\right.\right.$$
$$\left.\left.\leq 1-\rho\ or\ \phi\left(w_{u,v}-w_{\langle u,s\rangle,\langle v,t\rangle},q\right)\geq\rho\right\}\right|$$

$$\leq 1-\left(1-\frac{\sigma}{2}\right)+\frac{1}{mn}\left|\left\{(u,v)\in J_m\times I_n:\psi\left(w_{u,v}-w_{\langle u,s\rangle,\langle v,t\rangle},q\right)\right.\right.$$
$$\left.\left.\leq 1-\rho\ or\ \phi\left(w_{u,v}-w_{\langle u,s\rangle,\langle v,t\rangle},q\right)\geq\rho\right\}\right|$$

$$=\frac{\sigma}{2}+\frac{1}{mn}\left|\left\{(u,v)\in J_m\times I_n:\psi\left(w_{u,v}-w_{\langle u,s\rangle,\langle v,t\rangle},q\right)\leq 1-\rho\ or\ \phi\left(w_{u,v}-w_{\langle u,s\rangle,\langle v,t\rangle},q\right)\geq\rho\right\}\right|.$$

Then,

$$\left\{(m,n)\in\mathbb{N}\times\mathbb{N}:\frac{1}{mn}\left|\left\{u\leq m,v\leq n:\psi\left(w_{u,v}-w_{\langle u,s\rangle,\langle v,t\rangle},q\right)\leq 1-\rho\right.\right.\right.$$
$$\left.\left.\left.or\ \phi\left(w_{u,v}-w_{\langle u,s\rangle,\langle v,t\rangle},q\right)\geq\rho\right\}\right|\geq\sigma\right\}$$
$$\subset\left\{(m,n)\in\mathbb{N}\times\mathbb{N}:\frac{1}{mn}\left|\left\{(u,v)\in J_m\times I_n:\psi\left(w_{u,v}-w_{\langle u,s\rangle,\langle v,t\rangle},q\right)\leq 1-\rho\right.\right.\right.$$
$$\left.\left.\left.or\ \phi\left(w_{u,v}-w_{\langle u,s\rangle,\langle v,t\rangle},q\right)\geq\rho\right\}\right|\geq\frac{\sigma}{2}\right\}$$
$$\cup\{(1,1),(2,2),\ldots,(\check{m},\check{n})\}.$$

If $(\psi,\phi)-\lim w_{u,v}=w_{\langle u,s\rangle,\langle v,t\rangle}\left(ASt_{\lambda\mathscr{I}_2}\right)$, *then the set on the right-hand side belongs to* $\mathscr{I}_2$. *Hence,* $(\psi,\phi)-\lim_{u,v}=w_{\langle u,s\rangle,\langle v,t\rangle}\left(ASt_{\mathscr{I}_2}\right)$.

Below are some definitions of degree of convergence.

Definition 9.12 *A sequence* $(w_{u,v})$ *is named to be a arithmetic lacunary statistically convergent of order* α $(\alpha\in(0,1])$ *to* $w_{\langle u,s\rangle,\langle v,t\rangle}$ *on* $(Y,\psi,\phi,\diamond,*)$ *provided that for each* $\rho\in(0,1),q>0$, *and there are integers* s,t, *so that*

$$\lim_{m,n\to\infty}\frac{1}{\mathfrak{h}_{m,n}^{\alpha}}\left|\left\{(u,v)\in\mathscr{I}_{m,n}:\psi\left(w_{u,v}-w_{\langle u,s\rangle,\langle v,t\rangle},q\right)\leq 1-\rho\ or\ \phi\left(w_{u,v}-w_{\langle u,s\rangle,\langle v,t\rangle},q\right)\geq\rho\right\}\right|=0.$$

Symbolically, it is denoted as $(\psi,\phi)-\lim w_{u,v}=w_{\langle u,s\rangle,\langle v,t\rangle}\left(ASt_{\theta_2}{}^{\alpha}\right)$.

Definition 9.13 *A sequence* $(w_{u,v})$ *is called to be arithmetic* $\mathscr{I}_2$*-lacunary statistically convergent of order* α *to* $w_{\langle u,s\rangle,\langle v,t\rangle}$ *on* $(Y,\psi,\phi,\diamond,*)$ *provided that for each* $\rho\in(0,1),\sigma>0,q>0$, *and there are integers* s,t, *so that*

$$\left\{(m,n)\in\mathbb{N}\times\mathbb{N}:\frac{1}{\mathfrak{h}_{m,n}^{\alpha}}\left|\left\{(u,v)\in\mathscr{I}_{m,n}:\psi\left(w_{u,v}-w_{\langle u,s\rangle,\langle v,t\rangle},q\right)\leq 1-\rho\right.\right.\right.$$
$$\left.\left.\left.or\ \phi\left(w_{u,v}-w_{\langle u,s\rangle,\langle v,t\rangle},q\right)\geq\rho\right\}\right|\geq\sigma\right\}\in\mathscr{I}_2.$$

At that time, it is indicated $(\psi,\phi)-\lim w_{u,v}=w_{\langle u,s\rangle,\langle v,t\rangle}\left(ASt_{\mathscr{I}\theta_2}^{\alpha}\right)$.

Definition 9.14 *A sequence $(w_{u,v})$ is named to be arithmetic λ-statistically convergent of order α to $w_{\langle u,s\rangle,\langle v,t\rangle}$ on $(Y,\psi,\phi,\diamond,*)$ provided that for each $\rho\in(0,1), q>0$ and there are integers s,t, so that*

$$\lim_{m,n\to\infty}\frac{1}{\lambda_m^{\alpha}\mu_n^{\alpha}}\left|\left\{(u,v)\in J_m\times I_n:\psi\left(w_{u,v}-w_{\langle u,s\rangle,\langle v,t\rangle},q\right)\leq 1-\rho \text{ or } \phi\left(w_{u,v}-w_{\langle u,s\rangle,\langle v,t\rangle},q\right)\geq\rho\right\}\right|$$
$$=0.$$

Then, it is demonstrated $(\psi,\phi)-\lim w_{u,v}=w_{\langle u,s\rangle,\langle v,t\rangle}\left(ASt_{\lambda}^{\alpha}\right)$.

Definition 9.15 *The sequence $(w_{u,v})$ is called to be arithmetic arithmetic $\lambda\mathscr{I}$-statistically convergent of order α to $w_{\langle u,s\rangle,\langle v,t\rangle}$ on $(Y,\psi,\phi,\diamond,*)$ provided that for each $\rho\in(0,1),\sigma>0,q>0$ and there are integers s,t, such that*

$$\left\{(m,n)\in\mathbb{N}\times\mathbb{N}:\frac{1}{\lambda_m^{\alpha}\mu_n^{\alpha}}\left|\left\{(u,v)\in J_m\times I_n:\psi\left(w_{u,v}-w_{\langle u,s\rangle,\langle v,t\rangle},q\right)\leq 1-\rho\right.\right.\right.$$
$$\left.\left.\left.\text{or } \phi\left(w_{u,v}-w_{\langle u,s\rangle,\langle v,t\rangle},q\right)\geq\rho\right\}\right|\geq\sigma\right\}\in\mathscr{I}_2.$$

Then, it is indicated $(\psi,\phi)-\lim w_{u,v}=w_{\langle u,s\rangle,\langle v,t\rangle}\left(ASt_{\lambda\mathscr{I}_2}^{\alpha}\right)$.

The following inclusion relation for $(Y,\psi,\phi,\diamond,*)$ is valid if we show that all arithmetic $\mathscr{I}_2$ statistically convergent sequence spaces can be represented by $S\left(ASt_{\mathscr{I}_2}{}^{\alpha}\right)$.

Theorem 9.5 *Let $(Y,\psi,\phi,\diamond,*)$ be an IFNS, $\alpha,\mu\in(0,1]$ and $\alpha\leq\mu$. Then, $S\left(ASt_{\mathscr{I}\theta_2}^{\alpha}\right)\subset S\left(ASt_{\mathscr{I}\theta_2}^{\mu}\right)$.*

Proof *Let $\alpha,\mu\in(0,1]$ and $\alpha\leq\mu$. Then, we write*

$$\frac{1}{\mathfrak{h}_{m,n}^{\mu}}\left|\left\{(u,v)\in\mathscr{I}_{m,n}:\psi\left(w_{u,v}-w_{\langle u,s\rangle,\langle v,t\rangle},q\right)\leq 1-\rho \text{ or } \phi\left(w_{u,v}-w_{\langle u,s\rangle,\langle v,t\rangle},q\right)\geq\rho\right\}\right|$$
$$\leq\frac{1}{\mathfrak{h}_{m,n}^{\alpha}}\left|\left\{(u,v)\in\mathscr{I}_{m,n}:\psi\left(w_{u,v}-w_{\langle u,s\rangle,\langle v,t\rangle},q\right)\leq 1-\rho \text{ or } \phi\left(w_{u,v}-w_{\langle u,s\rangle,\langle v,t\rangle},q\right)\geq\rho\right\}\right|.$$

So, for each $\sigma>0$,

$$\left\{(m,n)\in\mathbb{N}\times\mathbb{N}:\frac{1}{\mathfrak{h}_{m,n}^{\mu}}\left|\left\{(u,v)\in\mathscr{I}_{m,n}:\psi\left(w_{u,v}-w_{\langle u,s\rangle,\langle v,t\rangle},q\right)\leq 1-\rho\right.\right.\right.$$
$$\left.\left.\left.\text{or } \phi\left(w_{u,v}-w_{\langle u,s\rangle,\langle v,t\rangle},q\right)\geq\rho\right\}\right|\geq\sigma\right\}$$
$$\subset\left\{(m,n)\in\mathbb{N}\times\mathbb{N}:\frac{1}{\mathfrak{h}_{m,n}^{\alpha}}\left|\left\{(u,v)\in\mathscr{I}_{m,n}:\psi\left(w_{u,v}-w_{\langle u,s\rangle,\langle v,t\rangle},q\right)\leq 1-\rho\right.\right.\right.$$
$$\left.\left.\left.\text{or } \phi\left(w_{u,v}-w_{\langle u,s\rangle,\langle v,t\rangle},q\right)\geq\rho\right\}\right|\geq\sigma\right\}.$$

Therefore, if an element belongs to $\mathscr{I}_2$ in the right-hand set, it also belongs to $\mathscr{I}_2$ in the left-hand set of this expression. Hence, $S\left(ASt_{\mathscr{I}\theta_2}^{\alpha}\right)\subset S\left(ASt_{\mathscr{I}\theta_2}^{\mu}\right)$.

9.4 CONCLUSIONS

In this research study, which holds significant importance among researchers in the field of functional analysis, an intuitionistic approach was employed to enhance the realism of decision-making processes when dealing with everyday problems. Various researchers have explored the concepts of ideals, lacunary convergence, and statistical convergence in IFNS. However, this particular study was undertaken due to the lack of investigation into arithmetic convergence within this context. The research aimed to examine the counterparts of λ-convergence in IFNS, and it was proposed that the concept of arithmetic convergence could be evaluated alongside λ-convergence in this regard. The study integrated this concept, which exhibits distinct properties across all three spaces utilized for decision processes with the aid of ideals, with the concept of arithmetic convergence.

This study was established to determine the properties of sequences that should be assessed in terms of arithmetic and statistical convergence in intuitionistic spaces, serving as a valuable resource for researchers who prioritize the type of convergence within their respective areas of work. A review of existing literature reveals that the deficiencies in the fuzzy structure have yet to be overcome, and the convergences associated with this type are typically examined in a classical sense, neglecting uncertainty and indecisive situations. Furthermore, no exploration of these studies' equivalents in the intuitionistic space has been conducted to date. Hence, the prepared paper represents a pioneering study that fills this gap.

Several definitions have been provided for different types of degree of convergence, and future studies can undertake comprehensive analyses from diverse perspectives to explore the interrelations between these concepts.

REFERENCES

1. Atanassov, K. T. (1986). Intuitionistic fuzzy sets, Fuzzy Sets Syst. **20**, 87–96.
2. Belen, C., Yildirim, M. (2012). On generalized statistical convergence of double sequences via ideals, Ann. Univ. Ferrara **58**(1), 11–20.
3. Das, P., Kostyrko, P., Wilczynski, W., Malik, P. (2008). $\mathscr{I}$ and $\mathscr{I}^*$-convergence of double sequences, Math. Slovaca **58**(5), 605–620.
4. Das, P., Savaş, E., Ghosal, S. Kr. (2011). On generalizations of certain summability methods using ideals, Appl. Math. Lett. **24**, 1509-1514.
5. Esi, A., Hazarika, B. (2013). λ-ideal convergence in intuitionistic fuzzy 2-normed linear space, J. Intell. Fuzzy Syst. **24**(4), 725–732.
6. Fast, H. (1951). Sur la convergence, Colloq. Math. **2**, 241–244.
7. George, V., Veeramani, P. (1994). On some results in fuzzy metric spaces, Fuzzy Sets Syst. **64**(3), 395–399.
8. Hazarika, B., Savaş, E. (2012). (λ,μ)-statistical convergence of double sequences in n-normed spaces, Note Mat. **32**, 101–114.
9. Kaleva, O., Seikkala, S. (1984). On fuzzy metric spaces, Fuzzy Sets Syst. **12**(3), 215–229.
10. Karakuş, S., Demirci, K., Duman, O. (2008). Statistical convergence on intuitionistic fuzzy normed spaces, Chaos Solitons Fractals **35**, 763–769.
11. Kişi, Ö. (2022). $\mathscr{I}$-lacunary arithmetic statistical convergence, J. Appl. Math. Inform., **40**(1-2), 327-339.
12. Konwar, N., Debnath, P. (2017). $\mathscr{I}_\lambda$-convergence in intuitionistic fuzzy n-normed linear space, Ann. Fuzzy Math. Inform. **13**(1), 91–107.
13. Kostyrko, P., Šalat, T., Wilczynski, W. (2000). $\mathscr{I}$-convergence, Real Anal. Exchange **26**(2), 669–686.
14. Kramosil, I., Michalek, J. (1975). Fuzzy metric and statistical metric spaces, Kybernetika **11**(5), 336–344.
15. Lael, F., Nourouzi, K. (2008). Some results on the IF-normed spaces, Chaos Solitons Fractals **37**, 931–939.
16. Mursaleen, M. (2000). λ-statistical convergence, Math. Slovaca **50**(1), 111–115.
17. Mursaleen, M., Edely, O.H.H. (2003). Statistical convergence of double sequences, J. Math. Anal. Appl. **288**, 223–231.

18. Mursaleen, M., Mohiuddine, S. A. (2009). Statistical convergence of double sequences in intuitionistic fuzzy normed spaces, Chaos Solitons Fractals **41**(5), 2414–2421.
19. Mursaleen, M., Mohiuddine, S. A., Edely, O. H. H. (2010). On the ideal convergence of double sequences in intuitionistic fuzzy normed spaces, Comput. Math. Appl. **59**, 603–611.
20. Park, J. H. (2004). Intuitionistic fuzzy metric spaces, Chaos Solitons Fractals **22**, 1039–1046.
21. Ruckle, W. H. (2012). Arithmetical summability, J. Math. Anal. Appl. **396**(2), 741–748.
22. Saadati, R., Park, J. H. (2006). On the intuitionistic fuzzy topological spaces, Chaos Solitons Fractals **27**, 331–344.
23. Savaş, E. (2008). On λ-statistically convergent double sequences of fuzzy numbers, J. Inequal. Appl. Art. ID 147827, 6 pp.
24. Savaş, E., Patterson, R. F. (2008). (λ,μ)-double sequence spaces via Orlicz function, J. Comput. Anal. Appl. **10**(1), 101–111.
25. Schweizer, B., Sklar, A. (1960). Statistical metric spaces, Pacific J. Math. **10**, 314–344.
26. Yaying, T., Hazarika, B. (2020). Lacunary arithmetic statistical convergence, Nat. Acad. Sci. Lett. **43**(6), 547–551.
27. Yaying, T., Hazarika, B. (2018). On lacunary arithmetic convergence, Proc. Jangjeon Math. Soc. **21**(3), 507–513.
28. Yaying, T., Hazarika, B. (2017). On arithmetical summability and multiplier sequences, Nat. Acad. Sci. Lett. **40**(1), 43–46.
29. Yaying, T., Hazarika, B. (2017). On arithmetic continuity, Bol. Soc. Parana Mater. **35**(1), 139–145.
30. Yaying, T., Hazarika, B. (2017). On arithmetic continuity in metric spaces, Afr. Mat. **28**, 985–989.
31. Zadeh, L. A. (1965). Fuzzy sets, Inf. Control. **8**(3), 338–353.

10 Fixed Point Approximation of Convex Contraction Mappings on a Nonlinear Domain

Ritika
Department of Mathematics, Pt. N. R. S. Government College,
M. D. University, Rohtak, Haryana, India

CONTENTS

Thakur et al. (2016) developed a three-step iterative process. For a convex contraction mapping of order ρ in CAT(0) spaces, we provide delta-convergence and strong convergence results in this study for the same iterative procedure. Some of the results in the existing literature are expanded upon by our studies.

10.1 INTRODUCTION

The area of mathematics known as fixed point theory is fascinating. It combines geometry, topology, and analysis. The fixed point theory has emerged as a very potent and crucial tool in the study of nonlinear processes over the past 50 years or so. It can be used in a wide variety of mathematical sciences fields. Fixed point theory is drawing young scholars from all over the world because of its significance. One of the active subfields of fixed point theory is metric fixed point theory, where geometric characteristics of the underlying space are crucial.

Let $\mathfrak{I}$ be a mapping from $\mathscr{H}$ to $\mathscr{H}$, where $\mathscr{H}$ is nonempty subset of a CAT(0) space $\mathscr{L}$. The point $\delta \in \mathscr{H}$ is said to be a fixed point of $\mathfrak{I}$ if $\mathfrak{I}\,\delta = \delta$. Here, we take the term $\wp(\mathfrak{I}) = \{\delta \in \mathscr{H} : \mathfrak{I}\,\delta = \delta\}$ as the set of fixed points of $\mathfrak{I}$.

10.2 PRELIMINARIES

Let's review a few definitions and findings from the literature that has already been written about this concept.

Given $(\mathscr{L}, \mathfrak{U})$ be a metric space. A mapping e from $[0, b] \subset \mathbb{R}$ to $\mathscr{L}$ such that $e(0) = \xi$, $e(b) = \eta$, $\mathfrak{U}(e(\mu), e(\mu')) = |\mu - \mu'|$ for all $\mu, \mu' \in [0, b]$, is known as a geodesic path between $\xi \in \mathscr{L}$ and

DOI: 10.1201/9781032634142-10

$\eta \in \mathscr{L}$ (or, more succinctly, a geodesic between ξ and η). In particular, $\mathfrak{U}(\xi, \eta)$ = b and e is an isometry. The image θ of e is called a geodesic (or metric) segment between ξ and η. This geodesic segment is indicated by $[\xi, \eta]$, when it is unique. If every two points of $\mathscr{L}$ are connected by a geodesic, the space $(\mathscr{L}, \mathfrak{U})$ is called a geodesic space. When there is precisely one geodesic joining ξ and η, for all $\xi, \eta \in \mathscr{L}$, then $\mathscr{L}$ is claimed to be uniquely geodesic. If $\mathscr{H}$ contains each and every geodesic segment connecting any two points in a subset $\mathscr{H} \subseteq \mathscr{L}$, then it is said to be convex. The three points ξ_1, ξ_2, ξ_3 in $\mathscr{L}$ and the geodesic segment between each pair of vertices make up a geodesic triangle $\Delta(\xi_1, \xi_2, \xi_3)$ in a geodesic metric space $(\mathscr{L}, \mathfrak{U})$. A comparison triangle for the geodesic triangle $\Delta(\xi_1, \xi_2, \xi_3)$ in $(\mathscr{L}, \mathfrak{U})$ is a triangle $\overline{\Delta}(\xi_1, \xi_2, \xi_3) = \Delta(\overline{\xi_1}, \overline{\xi_2}, \overline{\xi_3})$ in the Euclidean plane $\mathbb{R}^2$ such that $\mathfrak{U}_{\mathbb{R}^2}(\overline{\xi_i}, \overline{\xi_j}) = \mathfrak{U}(\xi_i, \xi_j)$ for $i, j \in \{1, 2, 3\}$.

If all geodesic triangles meet the under mentioned property, a geodesic space is said to be a CAT(0) space: Let $\overline{\Delta}$ be a comparison triangle for geodesic triangle Δ in $\mathscr{L}$. If $\mathfrak{U}(\xi, \eta) \leq \mathfrak{U}_{\mathbb{R}^2}(\overline{\xi}, \overline{\eta})$, then CAT(0) inequality is satisfied by the geodesic triangle Δ. This holds true for each $\xi, \eta \in \Delta$ and for all comparison points $\overline{\xi}, \overline{\eta} \in \overline{\Delta}$.

$$\mathfrak{U}(\xi, \eta_0)^2 \leq \frac{1}{2}\mathfrak{U}(\xi, \eta_1)^2 + \frac{1}{2}\mathfrak{U}(\xi, \eta_2)^2 - \frac{1}{4}\mathfrak{U}(\eta_1, \eta_2)^2 \text{(CN)}$$

The above inequality is implied by CAT(0) inequality, if ξ, η_1, η_2 are points in a CAT(0) space and if η_0 is the midpoint of the segment $[\eta_1, \eta_2]$.
In actuality, a geodesic space only qualifies as a CAT(0) space if it satisfies the (CN) inequality.

Lemma 10.1 *[6] In a CAT(0) space $\mathscr{L}$, this inequality holds*

$$\mathfrak{U}(\mu\eta \oplus (1-\mu)\xi, \tau) = \mu\mathfrak{U}(\eta, \tau) + (1-\mu)\mathfrak{U}(\xi, \tau)$$

for all $\xi, \eta, \tau \in \mathscr{L}$ and $\mu \in [0, 1]$.

Lemma 10.2 *[6] In a CAT(0) space $\mathscr{L}$, this inequality holds*

$$\mathfrak{U}^2((1-\mu)\xi \oplus \mu\eta, \tau) = (1-\mu)\mathfrak{U}^2(\xi, \tau) + \mu\mathfrak{U}^2(\eta, \tau) - \mu(1-\mu)\mathfrak{U}^2(\xi, \eta)$$

for all $\xi, \eta, \tau \in \mathscr{L}$ and $\mu \in [0, 1]$.

We take $r(\xi, \{\xi_\rho\}) = \lim sup_{\rho \to \infty} \mathfrak{U}(\xi, \xi_\rho)$ for $\xi \in \mathscr{L}$ (the CAT(0) space) and for bounded sequence $\{\xi_\rho\}$. The term $r(\{\xi_\rho\}) = inf\{ r(\xi, \{\xi_\rho\}) : \xi \in \mathscr{L}\}$ is defined as asymptotic radius of $\{\xi_\rho\}$ and $A(\{\xi_\rho\}) = \{ \xi \in \mathscr{L} : r(\xi, \{\xi_\rho\}) = r(\{\xi_\rho\})\}$is defined as the asymptotic center of $\{\xi_\rho\}$.

Remark 10.1 *[5] The asymptotic center $A(\{\xi_\rho\})$ of bounded sequence $\{\xi_\rho\}$ consists exactly one point in CAT(0) space.*

Definition 10.1 *[13] If each subsequence $\{\alpha_\rho\}$ of $\{\xi_\rho\}$ in a CAT(0) space $\mathscr{L}$ has a unique asymptotic centre ξ, then the sequence $\{\xi_\rho\}$ is said to Δ–converge to $\xi \in \mathscr{L}$. In this instance, we write $\Delta - \lim \xi_\rho = \xi$ and ξ refer to the Δ–limit of $\{\xi_\rho\}$.*

Noteworthy is the fact that $\lim sup_{\rho \to \infty} \mathfrak{U}(\xi_\rho, \xi) = \lim sup_{\rho \to \infty} \mathfrak{U}(\xi_\rho, \alpha)$, is the result of the uniqueness of the asymptotic center; given $\{\xi_\rho\} \subset \mathscr{L}$ such that $\{\xi_\rho\}$$\Delta$–converges to ξ and provided $\alpha \in \mathscr{L}$ with $\alpha \neq \xi$. Thus, the Opial property [1] is satisfied by every CAT(0) space.

Lemma 10.3 *[13] There is always a Δ-convergent subsequence for every bounded sequence, in complete CAT(0) space.*

Lemma 10.4 *[4] The asymptotic center of $\{\xi_\rho\}$ is in $\mathscr{H}$ if $\{\xi_\rho\}$ is a bounded sequence in $\mathscr{H}$ and $\mathscr{H}$ is convex and closed subset of a complete CAT(0) space.*

Definition 10.2 *[7] Let $(\mathscr{L}, \mathfrak{U})$ be a metric space. If there exist positive constants $\kappa_0, \kappa_1, \kappa_2, \ldots, \kappa_{\rho-1} \in (0,1)$ such that the following requirements are true:*

$1 \kappa_0 + \kappa_1 + \kappa_2 + \ldots + \kappa_{\rho-1} < 1.$
$2 \mathfrak{U}(\mathfrak{J}^{\rho}(\xi), \mathfrak{J}^{\rho}(\eta)) \leq \kappa_0 \mathfrak{U}(\xi, \eta) + \kappa_1 \mathfrak{U}(\mathfrak{J}(\xi), \mathfrak{J}(\eta)) + \kappa_2 \mathfrak{U}(\mathfrak{J}^2(\xi), \mathfrak{J}^2(\eta)) + \ldots +$
$3 \kappa_{\rho-1} \mathfrak{U}(\mathfrak{J}^{\rho-1}(\xi), \mathfrak{J}^{\rho-1}(\eta)).$

Then the continuous mapping $\mathfrak{J} : \mathscr{L} \to \mathscr{L}$ is said to be a convex contraction of order ρ.

Theorem 10.1 *[7] The mapping $\mathfrak{J}$ has a unique fixed point δ, that is, $\wp(\mathfrak{J}) = \{\delta\}$, if $\mathfrak{J} : \mathscr{L} \to \mathscr{L}$ be a convex contraction of order ρ on a complete metric space $\mathscr{L}$.*

Definition 10.3 *[21] Make $\mathscr{L}$, a set that is non – empty. If there exists $\delta \in \mathscr{L}$ such that $\wp(\mathfrak{J}^{\rho}) = \{\delta\}$ for all $\rho \in \mathbb{N}$, the map $\mathfrak{J} : \mathscr{L} \to \mathscr{L}$ is said to be a Bessaga mapping.*

Theorem 10.2 *[21] The mapping $\mathfrak{J}$ is a Bessaga mapping if $\mathfrak{J} : \mathscr{L} \to \mathscr{L}$ is a convex contraction of order ρ on a metric space $\mathscr{L}$.*

Lemma 10.5 *[21] Let a convex contraction mapping $\mathfrak{J}$ of order ρ defined from $\mathscr{H}$ to $\mathscr{H}$, where $\mathscr{H}$ is non–empty convex and closed subset of a complete CAT(0) space $\mathscr{L}$. If a sequence $\{\xi_\rho\}$ delta – converges to ξ and $\mathfrak{U}(\xi_\rho, \mathfrak{J}(\xi_\rho)) \to 0$, then $\xi \in \mathscr{H}$ and $\mathfrak{J}(\xi) = \xi$.*

Thakur et al. [20] proposed an iterative process and showed that it converges substantially more quickly than any other iterative process that has been documented so far. The following sequence $\{\xi_\rho\}$ describes Thakur's [20] iterative process:

$$\begin{aligned} \xi_{\rho+1} &= (1-\sigma_\rho)\mathfrak{J}\,\tau_\rho + \sigma_\rho \mathfrak{J}\eta_\rho, \\ \eta_\rho &= (1-\omega_\rho)\tau_\rho + \omega_\rho \mathfrak{J}\tau_\rho, \\ \tau_\rho &= (1-\vartheta_\rho)\xi_\rho + \vartheta_\rho \mathfrak{J}\xi_\rho, \end{aligned}$$

where $\{\sigma_\rho\}$, $\{\omega_\rho\}$, $\{\vartheta_\rho\}$ are positive valued sequences in [0, 1] with $\sum_{\rho=0}^{\infty} \sigma_\rho = \infty$.

As the aforementioned iteration procedure converges significantly more quickly than others found in the literature, several scholars have shown interest in it [10, 17, 18, 19]. For generalized nonexpansive mappings in CAT(0) space settings, Ritika [15] adapted the aforementioned iteration procedure in 2018. Ritika [14] changed this iteration procedure for common fixed points of generalized quasi-contractive type operators. In 2020, Ritika [16] made additional modifications to this iteration procedure for generalized asymptotically quasi–nonexpansive in CAT(0) spaces. Additionally, Khan and Ritika [9] found common fixed points of mean nonexpansive mappings in CAT(0) spaces using the same iterative process in 2020.

This idea of Thakur et al. [20] iterative procedure in CAT(0) spaces is now modified to account for convex contraction mappings of order ρ.

Let a convex contraction mapping $\mathfrak{J}$ of order ρ defined from $\mathscr{H}$ to $\mathscr{H}$, where $\mathscr{H}$ is non–empty convex and closed subset of a complete CAT(0) space $\mathscr{L}$ with $\wp(\mathfrak{J}) \neq \emptyset$. Suppose that $\{\xi_\rho\}$ is a sequence generated iteratively by

$$\begin{aligned} \xi_{\rho+1} &= (1-\sigma_\rho)\mathfrak{J}^{\rho}\tau_\rho \oplus \sigma_\rho \mathfrak{J}^{\rho}\eta_\rho, \\ \eta_\rho &= (1-\omega_\rho)\tau_\rho \oplus \omega_\rho \mathfrak{J}^{\rho}\tau_\rho, \\ \tau_\rho &= (1-\vartheta_\rho)\xi_\rho \oplus \vartheta_\rho \mathfrak{J}^{\rho}\xi_\rho, \end{aligned} \tag{1}$$

where $\{\sigma_\rho\}$, $\{\omega_\rho\}$, $\{\vartheta_\rho\}$ are positive valued sequences in $[0,1]$.

In this article, we investigate the three-step iterative method proposed by Thakur et al. [20] for convex contraction mappings of order ρ and establish several convergence findings, such as the delta-convergence result and the strong convergence result. And every convex contraction of order ρ is Bessaga mapping. So our results also hold for Bessaga mappings in CAT(0) spaces.

10.3 MAIN RESULTS

For the three-step iterative process proposed here in (1) in the CAT(0) spaces, we demonstrate the delta-convergence and strong convergence theorems in this section.

Lemma 10.6 *Let a convex contraction mapping $\mathfrak{I}$ of order ρ defined from $\mathscr{H}$ to $\mathscr{H}$, where $\mathscr{H}$ is non–empty convex and closed subset of a complete CAT(0) space $\mathscr{L}$ and $\{\xi_\rho\}$ is an iterative process defined above in* (1). *Then* $\lim_{\rho\to\infty}\mathfrak{U}(\xi_\rho, \delta)$ *exists for* $\delta \in \wp(\mathfrak{I})$.

Proof Let $\delta \in \wp(\mathfrak{I})$.

$$
\begin{aligned}
&\mathfrak{U}(\xi_{\rho+1},\delta) = \mathfrak{U}((1-\sigma_\rho)\mathfrak{I}^{\rho}\tau_\rho \oplus \sigma_\rho\mathfrak{I}^{\rho}\eta_\rho,\delta)\\
&\le (1-\sigma_\rho)\mathfrak{U}(\mathfrak{I}^{\rho}\tau_\rho,\delta)+\sigma_\rho\mathfrak{U}(\mathfrak{I}^{\rho}\eta_\rho,\delta)\\
&= (1-\sigma_\rho)\mathfrak{U}(\mathfrak{I}^{\rho}\tau_\rho,\mathfrak{I}^{\rho}\delta)+\sigma_\rho\mathfrak{U}(\mathfrak{I}^{\rho}\eta_\rho,\mathfrak{I}^{\rho}\delta)\\
&\le (1-\sigma_\rho)[\kappa_0\mathfrak{U}(\tau_\rho,\delta)+\kappa_1\mathfrak{U}(\mathfrak{I}\tau_\rho,\mathfrak{I}\delta)+\kappa_2\mathfrak{U}(\mathfrak{I}^{2}\tau_\rho,\mathfrak{I}^{2}\delta)+\ldots+\kappa_{\rho-1}\mathfrak{U}(\mathfrak{I}^{\rho-1}\tau_\rho,\mathfrak{I}^{\rho-1}\delta)]\\
&+\sigma_\rho[\kappa_0\mathfrak{U}(\eta_\rho,\delta)+\kappa_1\mathfrak{U}(\mathfrak{I}\eta_\rho,\mathfrak{I}\delta)+\kappa_2\mathfrak{U}(\mathfrak{I}^{2}\eta_\rho,\mathfrak{I}^{2}\delta)+\ldots+\kappa_{\rho-1}\mathfrak{U}(\mathfrak{I}^{\rho-1}\eta_\rho,\mathfrak{I}^{\rho-1}\delta)]\\
&\le (1-\sigma_\rho)[\kappa_0\mathfrak{U}(\tau_\rho,\delta)+\kappa_1\kappa_0\mathfrak{U}(\tau_\rho,\delta)+\kappa_2\{\kappa_0\mathfrak{U}(\tau_\rho,\delta)+\kappa_1\kappa_0\mathfrak{U}(\tau_\rho,\delta)\}\\
&+\ldots+\kappa_{\rho-1}\{\kappa_0\mathfrak{U}(\tau_\rho,\delta)+\kappa_1\kappa_0\mathfrak{U}(\tau_\rho,\delta)+\kappa_2\{\kappa_0\mathfrak{U}(\tau_\rho,\delta)+\kappa_1\kappa_0\mathfrak{U}(\tau_\rho,\delta)\}\\
&+\kappa_3(\kappa_0\mathfrak{U}(\tau_\rho,\delta)+\kappa_1\kappa_0\mathfrak{U}(\tau_\rho,\delta)+\kappa_2\{\kappa_0\mathfrak{U}(\tau_\rho,\delta)+\kappa_1\kappa_0\mathfrak{U}(\tau_\rho,\delta)\})+\ldots\}]\\
&+\sigma_\rho[\kappa_0\mathfrak{U}(\eta_\rho,\delta)+\kappa_1\kappa_0\mathfrak{U}(\eta_\rho,\delta)+\kappa_2\{\kappa_0\mathfrak{U}(\eta_\rho,\delta)+\kappa_1\kappa_0\mathfrak{U}(\eta_\rho,\delta)\}\\
&+\ldots+\kappa_{\rho-1}\{\kappa_0\mathfrak{U}(\eta_\rho,\delta)+\kappa_1\kappa_0\mathfrak{U}(\eta_\rho,\delta)+\kappa_2\{\kappa_0\mathfrak{U}(\eta_\rho,\delta)+\kappa_1\kappa_0\mathfrak{U}(\eta_\rho,\delta)\}\\
&+\kappa_3(\kappa_0\mathfrak{U}(\eta_\rho,\delta)+\kappa_1\kappa_0\mathfrak{U}(\eta_\rho,\delta)+\kappa_2\{\kappa_0\mathfrak{U}(\eta_\rho,\delta)+\kappa_1\kappa_0\mathfrak{U}(\eta_\rho,\delta)\})+\ldots\}]\\
&= (1-\sigma_\rho)\mathfrak{U}(\tau_\rho,\delta)[\kappa_0+\kappa_1\kappa_0+\kappa_2(\kappa_0+\kappa_1\kappa_0)+\kappa_3(\kappa_0+\kappa_1\kappa_0+\kappa_2\{\kappa_0+\kappa_1\kappa_0\})\\
&+\ldots+\kappa_{\rho-1}\{\kappa_0+\kappa_1\kappa_0+\kappa_2\{\kappa_0+\kappa_1\kappa_0\}+\kappa_3(\kappa_0+\kappa_1\kappa_0+\kappa_2\{\kappa_0+\kappa_1\kappa_0\})+\ldots\}]\\
&+\sigma_\rho\mathfrak{U}(\eta_\rho,\delta)[\kappa_0+\kappa_1\kappa_0+\kappa_2(\kappa_0+\kappa_1\kappa_0)+\kappa_3(\kappa_0+\kappa_1\kappa_0+\kappa_2\{\kappa_0+\kappa_1\kappa_0\})\\
&+\ldots+\kappa_{\rho-1}\{\kappa_0+\kappa_1\kappa_0+\kappa_2\{\kappa_0+\kappa_1\kappa_0\}+\kappa_3(\kappa_0+\kappa_1\kappa_0+\kappa_2\{\kappa_0+\kappa_1\kappa_0\})+\ldots\}]
\end{aligned}
$$

$$
\begin{aligned}
&= (1-\sigma_\rho)\kappa_0\mathfrak{U}(\tau_\rho,\delta)[1+\kappa_1+\kappa_2(1+\kappa_1)+\kappa_3(1+\kappa_1+\kappa_2\{1+\kappa_1\})\\
&+\ldots+\kappa_{\rho-1}\{1+\kappa_1+\kappa_2\{1+\kappa_1\}+\kappa_3(1+\kappa_1+\kappa_2\{1+\kappa_1\})+\ldots\}]\\
&+\sigma_\rho\kappa_0\mathfrak{U}(\eta_\rho,\delta)[1+\kappa_1+\kappa_2(1+\kappa_1)+\kappa_3(1+\kappa_1+\kappa_2\{1+\kappa_1\})\\
&+\ldots+\kappa_{\rho-1}\{1+\kappa_1+\kappa_2\{1+\kappa_1\}+\kappa_3(1+\kappa_1+\kappa_2\{1+\kappa_1\})+\ldots\}]\\
&= (1-\sigma_\rho)\kappa_0(1+\kappa_1)\mathfrak{U}(\tau_\rho,\delta)[1+\kappa_2+\kappa_3(1+\kappa_2)+\ldots+\kappa_{\rho-1}\{1+\kappa_2+\kappa_3(1+\kappa_2)+\ldots\}]\\
&+\sigma_\rho\kappa_0(1+\kappa_1)\mathfrak{U}(\eta_\rho,\delta)[1+\kappa_2+\kappa_3(1+\kappa_2)+\ldots+\kappa_{\rho-1}\{1+\kappa_2+\kappa_3(1+\kappa_2)+\ldots\}]\\
&= (1-\sigma_\rho)\kappa_0(1+\kappa_1)(1+\kappa_2)\mathfrak{U}(\tau_\rho,\delta)[1+\kappa_3+\ldots+\kappa_{\rho-1}\{1+\kappa_3+\ldots\}]\\
&+\sigma_\rho\kappa_0(1+\kappa_1)(1+\kappa_2)\mathfrak{U}(\eta_\rho,\delta)[1+\kappa_3+\ldots+\kappa_{\rho-1}\{1+\kappa_3+\ldots\}]\\
&= (1-\sigma_\rho)\kappa_0(1+\kappa_1)(1+\kappa_2)(1+\kappa_3)\ldots(1+\kappa_{\rho-1})\mathfrak{U}(\tau_\rho,\delta)\\
&+\sigma_\rho\kappa_0(1+\kappa_1)(1+\kappa_2)(1+\kappa_3)\ldots(1+\kappa_{\rho-1})\mathfrak{U}(\eta_\rho,\delta)\\
&= \kappa_0(1+\kappa_1)(1+\kappa_2)(1+\kappa_3)\ldots(1+\kappa_{\rho-1})[(1-\sigma_\rho)\mathfrak{U}(\tau_\rho,\delta)+\sigma_\rho\mathfrak{U}(\eta_\rho,\delta)]
\end{aligned}
$$

$$
\mathfrak{U}(\xi_{\rho+1},\delta) = \mathscr{K}[(1-\sigma_\rho)\mathfrak{U}(\tau_\rho,\delta)+\sigma_\rho\mathfrak{U}(\eta_\rho,\delta)], \tag{2}
$$

where $\mathscr{K} = \kappa_0(1+\kappa_1)(1+\kappa_2)(1+\kappa_3)\ldots(1+\kappa_{\rho-1})$.

$$\begin{aligned}
&\mathfrak{U}(\eta_\rho,\delta) = \mathfrak{U}((1-\omega_\rho)\tau_\rho \oplus \omega_\rho \mathfrak{I}^{\rho}\tau_\rho,\delta)\\
&\leq (1-\omega_\rho)\mathfrak{U}(\tau_\rho,\delta) + \omega_\rho\mathfrak{U}(\mathfrak{I}^{\rho}\tau_\rho,\delta) = (1-\omega_\rho)\mathfrak{U}(\tau_\rho,\delta) + \omega_\rho\mathfrak{U}(\mathfrak{I}^{\rho}\tau_\rho,\mathfrak{I}^{\rho}\delta)\\
&\leq (1-\omega_\rho)\mathfrak{U}(\tau_\rho,\delta) + \omega_\rho[\kappa_0\mathfrak{U}(\tau_\rho,\delta) + \kappa_1\mathfrak{U}(\mathfrak{I}\tau_\rho,\mathfrak{I}\delta) + \kappa_2\mathfrak{U}(\mathfrak{I}^2\tau_\rho,\mathfrak{I}^2\delta) + \ldots +\\
&\kappa_{\rho-1}\mathfrak{U}(\mathfrak{I}^{\rho-1}\tau_\rho,\mathfrak{I}^{\rho-1}\delta)]\\
&\leq (1-\omega_\rho)\mathfrak{U}(\tau_\rho,\delta) + \omega_\rho[\kappa_0\mathfrak{U}(\tau_\rho,\delta) + \kappa_1\kappa_0\mathfrak{U}(\tau_\rho,\delta) + \kappa_2\{\kappa_0\mathfrak{U}(\tau_\rho,\delta) + \kappa_1\kappa_0\mathfrak{U}(\tau_\rho,\delta)\}\\
&+ \ldots + \kappa_{\rho-1}\{\kappa_0\mathfrak{U}(\tau_\rho,\delta) + \kappa_1\kappa_0\mathfrak{U}(\tau_\rho,\delta) + \kappa_2\{\kappa_0\mathfrak{U}(\tau_\rho,\delta) + \kappa_1\kappa_0\mathfrak{U}(\tau_\rho,\delta)\}\\
&+ \kappa_3(\kappa_0\mathfrak{U}(\tau_\rho,\delta) + \kappa_1\kappa_0\mathfrak{U}(\tau_\rho,\delta) + \kappa_2\{\kappa_0\mathfrak{U}(\tau_\rho,\delta) + \kappa_1\kappa_0\mathfrak{U}(\tau_\rho,\delta)\}) + \ldots\}]\\
&= (1-\omega_\rho)\mathfrak{U}(\tau_\rho,\delta) + \omega_\rho\mathfrak{U}(\tau_\rho,\delta)[\kappa_0 + \kappa_1\kappa_0 + \kappa_2(\kappa_0 + \kappa_1\kappa_0)\\
&+ \kappa_3(\kappa_0 + \kappa_1\kappa_0 + \kappa_2\{\kappa_0 + \kappa_1\kappa_0\})\\
&+ \ldots + \kappa_{\rho-1}\{\kappa_0 + \kappa_1\kappa_0 + \kappa_2\{\kappa_0 + \kappa_1\kappa_0\} + \kappa_3(\kappa_0 + \kappa_1\kappa_0 + \kappa_2\{\kappa_0 + \kappa_1\kappa_0\}) + \ldots\}]\\
&= (1-\omega_\rho)\mathfrak{U}(\tau_\rho,\delta) + \omega_\rho\kappa_0\mathfrak{U}(\tau_\rho,\delta)[1 + \kappa_1 + \kappa_2(1+\kappa_1) + \kappa_3(1 + \kappa_1 + \kappa_2\{1+\kappa_1\})\\
&+ \ldots + \kappa_{\rho-1}\{1+\kappa_1 + \kappa_2\{1+\kappa_1\} + \kappa_3(1+\kappa_1 + \kappa_2\{1+\kappa_1\}) + \ldots\}]\\
&= (1-\omega_\rho)\mathfrak{U}(\tau_\rho,\delta) + \omega_\rho\kappa_0(1+\kappa_1)\mathfrak{U}(\tau_\rho,\delta)[1 + \kappa_2 + \kappa_3(1+\kappa_2) + \ldots +\\
&\kappa_{\rho-1}\{1+\kappa_2 + \kappa_3(1+\kappa_2) + \ldots\}]\\
&= (1-\omega_\rho)\mathfrak{U}(\tau_\rho,\delta) + \omega_\rho\kappa_0(1+\kappa_1)(1+\kappa_2)\mathfrak{U}(\tau_\rho,\delta)[1 + \kappa_3 + \ldots + \kappa_{\rho-1}\{1+\kappa_3 + \ldots\}]\\
&= (1-\omega_\rho)\mathfrak{U}(\tau_\rho,\delta) + \omega_\rho\kappa_0(1+\kappa_1)(1+\kappa_2)(1+\kappa_3)\ldots(1+\kappa_{\rho-1})\mathfrak{U}(\tau_\rho,\delta)\\
&= \mathfrak{U}(\tau_\rho,\delta)[1-\omega_\rho + \omega_\rho\kappa_0(1+\kappa_1)(1+\kappa_2)(1+\kappa_3)\ldots(1+\kappa_{\rho-1})]\\
&= \mathfrak{U}(\tau_\rho,\delta)[1-\omega_\rho + \omega_\rho\mathscr{K}],
\end{aligned}$$

where $\mathscr{K} = \kappa_0(1+\kappa_1)(1+\kappa_2)(1+\kappa_3)\ldots(1+\kappa_{\rho-1})$.

$$\mathfrak{U}(\eta_\rho,\delta) = [1-\omega_\rho(1-\mathscr{K})]\mathfrak{U}(\tau_\rho,\delta) \tag{3}$$

$$\begin{aligned}
&\mathfrak{U}(\tau_\rho,\delta) = \mathfrak{U}((1-\vartheta_\rho)\xi_\rho \oplus \vartheta_\rho\mathfrak{I}^{\rho}\xi_\rho,\delta)\\
&\leq (1-\vartheta_\rho)\mathfrak{U}(\xi_\rho,\delta) + \vartheta_\rho\mathfrak{U}(\mathfrak{I}^{\rho}\xi_\rho,\delta)\\
&= (1-\vartheta_\rho)\mathfrak{U}(\xi_\rho,\delta) + \vartheta_\rho\mathfrak{U}(\mathfrak{I}^{\rho}\xi_\rho,\mathfrak{I}^{\rho}\delta)\\
&\leq (1-\vartheta_\rho)\mathfrak{U}(\xi_\rho,\delta) + \vartheta_\rho[\kappa_0\mathfrak{U}(\xi_\rho,\delta) + \kappa_1\mathfrak{U}(\mathfrak{I}\xi_\rho,\mathfrak{I}\delta) + \kappa_2\mathfrak{U}(\mathfrak{I}^2\xi_\rho,\mathfrak{I}^2\delta) + \ldots +\\
&\kappa_{\rho-1}\mathfrak{U}(\mathfrak{I}^{\rho-1}\xi_\rho,\mathfrak{I}^{\rho-1}\delta)]\\
&\leq (1-\vartheta_\rho)\mathfrak{U}(\xi_\rho,\delta) + \vartheta_\rho[\kappa_0\mathfrak{U}(\xi_\rho,\delta) + \kappa_1\kappa_0\mathfrak{U}(\xi_\rho,\delta) + \kappa_2\{\kappa_0\mathfrak{U}(\xi_\rho,\delta) + \kappa_1\kappa_0\mathfrak{U}(\xi_\rho,\delta)\}\\
&+ \ldots + \kappa_{\rho-1}\{\kappa_0\mathfrak{U}(\xi_\rho,\delta) + \kappa_1\kappa_0\mathfrak{U}(\xi_\rho,\delta) + \kappa_2\{\kappa_0\mathfrak{U}(\xi_\rho,\delta) + \kappa_1\kappa_0\mathfrak{U}(\xi_\rho,\delta)\} +\\
&\kappa_3\left(\kappa_0\mathfrak{U}(\xi_\rho,\delta) + \kappa_1\kappa_0\mathfrak{U}\left(\xi_\rho,\delta\right) + \kappa_2\{\kappa_0\mathfrak{U}(\xi_\rho,\delta) + \kappa_1\kappa_0\mathfrak{U}(\xi_\rho,\delta)\}\right) + \ldots\}]\\
&= (1-\vartheta_\rho)\mathfrak{U}(\xi_\rho,\delta) + \vartheta_\rho\mathfrak{U}(\xi_\rho,\delta)[\kappa_0 + \kappa_1\kappa_0 + \kappa_2(\kappa_0 + \kappa_1\kappa_0)\\
&+ \kappa_3(\kappa_0 + \kappa_1\kappa_0 + \kappa_2\{\kappa_0 + \kappa_1\kappa_0\})\\
&+ \ldots + \kappa_{\rho-1}\{\kappa_0 + \kappa_1\kappa_0 + \kappa_2\{\kappa_0 + \kappa_1\kappa_0\} + \kappa_3(\kappa_0 + \kappa_1\kappa_0 + \kappa_2\{\kappa_0 + \kappa_1\kappa_0\}) + \ldots\}]\\
&= (1-\vartheta_\rho)\mathfrak{U}(\xi_\rho,\delta) + \vartheta_\rho\kappa_0\mathfrak{U}(\xi_\rho,\delta)[1 + \kappa_1 + \kappa_2(1+\kappa_1) + \kappa_3(1 + \kappa_1 + \kappa_2\{1+\kappa_1\})\\
&+ \ldots + \kappa_{\rho-1}\{1+\kappa_1 + \kappa_2\{1+\kappa_1\} + \kappa_3(1+\kappa_1 + \kappa_2\{1+\kappa_1\}) + \ldots\}]\\
&= (1-\vartheta_\rho)\mathfrak{U}(\xi_\rho,\delta) + \vartheta_\rho\kappa_0(1+\kappa_1)\mathfrak{U}(\xi_\rho,\delta)[1 + \kappa_2 + \kappa_3(1+\kappa_2) + \ldots +
\end{aligned}$$

$$\kappa_{\rho-1}\{1+\kappa_2+\kappa_3(1+\kappa_2)+\ldots\}]$$
$$=(1-\vartheta_\rho)\mathfrak{U}(\xi_\rho,\delta)+\vartheta_\rho\kappa_0(1+\kappa_1)(1+\kappa_2)\mathfrak{U}(\xi_\rho,\delta)[1+\kappa_3+\ldots+\kappa_{\rho-1}\{1+\kappa_3+\ldots\}]$$
$$=(1-\vartheta_\rho)\mathfrak{U}(\xi_\rho,\delta)+\vartheta_\rho\kappa_0(1+\kappa_1)(1+\kappa_2)(1+\kappa_3)\ldots(1+\kappa_{\rho-1})\mathfrak{U}(\xi_\rho,\delta)$$
$$=\mathfrak{U}(\xi_\rho,\delta)[1-\vartheta_\rho+\vartheta_\rho\kappa_0(1+\kappa_1)(1+\kappa_2)(1+\kappa_3)\ldots(1+\kappa_{\rho-1})]$$
$$=\mathfrak{U}(\xi_\rho,\delta)[1-\vartheta_\rho+\vartheta_\rho\mathscr{K}],$$

where $\mathscr{K}=\kappa_0(1+\kappa_1)(1+\kappa_2)(1+\kappa_3)\ldots(1+\kappa_{\rho-1})$.

$$\mathfrak{U}(\tau_\rho,\delta)=[1-\vartheta_\rho(1-\mathscr{K})]\mathfrak{U}(\xi_\rho,\delta) \tag{4}$$

Putting the value of $\mathfrak{U}(\tau_\rho,\delta)$ from equation (4) into the equation (3), we get

$$\mathfrak{U}(\eta_\rho,\delta)\leq[1-\omega_\rho(1-\mathscr{K})][1-\vartheta_\rho(1-\mathscr{K})]\mathfrak{U}(\xi_\rho,\delta) \tag{5}$$

Putting the values of $\mathfrak{U}(\eta_\rho,\delta)$ and $\mathfrak{U}(\tau_\rho,\delta)$ from equations (5) and (4) into the equation (2), we get

$$\begin{aligned}\mathfrak{U}(\xi_{\rho+1},\delta)&\leq\mathscr{K}[(1-\sigma_\rho)\mathfrak{U}(\tau_\rho,\delta)+\sigma_\rho\mathfrak{U}(\eta_\rho,\delta)]\\&\leq\mathscr{K}[(1-\sigma_\rho)[1-\vartheta_\rho(1-\mathscr{K})]\mathfrak{U}(\xi_\rho,\delta)+\sigma_\rho[1-\omega_\rho(1-\mathscr{K})]\\&\quad[1-\vartheta_\rho(1-\mathscr{K})]\mathfrak{U}(\xi_\rho,\delta)].\end{aligned}$$

Since we have $\mathscr{K}<1$, so,

$$\mathfrak{U}(\xi_{\rho+1},\delta)\leq(1-\sigma_\rho)\mathfrak{U}(\xi_\rho,\delta)+\sigma_\rho\mathfrak{U}(\xi_\rho,\delta)=\mathfrak{U}(\xi_\rho,\delta),$$

which shows that the sequence $\{\mathfrak{U}(\xi_\rho,\delta)\}$ is a decreasing sequence and bounded below and hence $\lim_{\rho\to\infty}\mathfrak{U}(\xi_\rho,\delta)$ exists for $\delta\in\wp(\mathfrak{J})$.

Lemma 10.7 *Let a convex contraction mapping $\mathfrak{J}$ of order ρ defined from $\mathscr{H}$ to $\mathscr{H}$, where $\mathscr{H}$ is non-empty convex and closed subset of a complete CAT(0) space $\mathscr{L}$ and $\{\xi_\rho\}$ is an iterative process defined above in* (1) *with* $0<a\leq\sigma_\rho,\omega_\rho,\vartheta_\rho\leq b<1$. *Then* $\lim_{\rho\to\infty}\mathfrak{U}(\mathfrak{J}\xi_\rho,\xi_\rho)=0$.

Proof *Let $\delta\in\wp(\mathfrak{J})$ and $\mathfrak{J}$ be a convex contraction of order ρ.*
By previous Lemma 10.1, we get $\mathfrak{U}(\xi_{\rho+1},\delta)\leq(1-\sigma_\rho)\mathfrak{U}(\tau_\rho,\delta)+\sigma_\rho\mathfrak{U}(\eta_\rho,\delta)$.
Also $\mathfrak{U}(\eta_\rho,\delta)\leq\mathfrak{U}(\tau_\rho,\delta)$ *and* $\mathfrak{U}(\tau_\rho,\delta)\leq\mathfrak{U}(\xi_\rho,\delta)$.
So, $\mathfrak{U}(\xi_{\rho+1},\delta)\leq\mathfrak{U}(\tau_\rho,\delta)\leq\mathfrak{U}(\xi_\rho,\delta)$.
By Lemma 10.1, we know that $\lim_{\rho\to\infty}\mathfrak{U}(\xi_\rho,\delta)$ *exists for* $\delta\in\wp(\mathfrak{J})$.
Let

$$\lim_{\rho\to\infty}\mathfrak{U}(\xi_\rho,\delta)=\ell. \tag{6}$$

$\lim\, inf_{\rho\to\infty}\mathfrak{U}(\xi_{\rho+1},\delta)\leq\lim\, inf_{\rho\to\infty}\mathfrak{U}(\tau_\rho,\delta)$ *implies*

$$\ell\leq\lim\, inf_{\rho\to\infty}\mathfrak{U}(\tau_\rho,\delta). \tag{7}$$

On the other hand, $\mathfrak{U}(\tau_\rho,\delta)\leq\mathfrak{U}(\xi_\rho,\delta)$ *implies that*

$$\lim\, sup_{\rho\to\infty}\mathfrak{U}(\tau_\rho,\delta)\leq\ell. \tag{8}$$

From equations (7) *and* (8), *we have*

$$\lim_{\rho\to\infty}\mathfrak{U}(\tau_\rho,\delta)=\ell. \tag{9}$$

Now,

$$\begin{aligned}
&\mathfrak{U}^2(\tau_\rho,\delta)=\mathfrak{U}^2((1-\vartheta_\rho)\xi_\rho\oplus\vartheta_\rho\mathfrak{J}^{\rho}\xi_\rho,\delta)\\
&\le(1-\vartheta_\rho)\mathfrak{U}^2(\xi_\rho,\delta)+\vartheta_\rho\mathfrak{U}^2(\mathfrak{J}^{\rho}\xi_\rho,\delta)-\vartheta_\rho(1-\vartheta_\rho)\mathfrak{U}^2(\xi_\rho,\mathfrak{J}^{\rho}\xi_\rho)\\
&\le(1-\vartheta_\rho)\mathfrak{U}^2(\xi_\rho,\delta)+\vartheta_\rho\mathfrak{U}^2(\mathfrak{J}^{\rho}\xi_\rho,\mathfrak{J}^{\rho}\,\delta)-\vartheta_\rho(1-\vartheta_\rho)\mathfrak{U}^2(\xi_\rho,\mathfrak{J}^{\rho}\xi_\rho)\\
&\le(1-\vartheta_\rho)\mathfrak{U}^2(\xi_\rho,\delta)+\vartheta_\rho[\kappa_0\mathfrak{U}(\xi_\rho,\delta)+\kappa_1\mathfrak{U}(\mathfrak{J}\xi_\rho,\mathfrak{J}\delta)+\kappa_2\mathfrak{U}(\mathfrak{J}^{2}\xi_\rho,\mathfrak{J}^{2}\delta)\\
&+\ldots+\kappa_{\rho-1}\mathfrak{U}(\mathfrak{J}^{\rho-1}\xi_\rho,\mathfrak{J}^{\rho-1}\delta)]^2-\vartheta_\rho(1-\vartheta_\rho)\mathfrak{U}^2(\xi_\rho,\mathfrak{J}^{\rho}\xi_\rho)\\
&\le(1-\vartheta_\rho)\mathfrak{U}^2(\xi_\rho,\delta)+\vartheta_\rho[\kappa_0\mathfrak{U}(\xi_\rho,\delta)+\kappa_1\kappa_0\mathfrak{U}(\xi_\rho,\delta)+\kappa_2\{\kappa_0\mathfrak{U}(\xi_\rho,\delta)+\kappa_1\kappa_0\mathfrak{U}(\xi_\rho,\delta)\}\\
&+\ldots+\kappa_{\rho-1}\{\kappa_0\,\mathfrak{U}(\,\xi_\rho\,,\,\delta)+\,\kappa_1\,\kappa_0\,\mathfrak{U}(\,\xi_\rho\,,\,\delta)+\,\kappa_2\,\{\,\kappa_0\,\mathfrak{U}(\,\xi_\rho\,,\,\delta)+\,\kappa_1\,\kappa_0\,\mathfrak{U}(\,\xi_\rho\,,\,\delta)\}+\\
&\kappa_3\,(\kappa_0\,\mathfrak{U}(\,\xi_\rho\,,\,\delta)+\,\kappa_1\,\kappa_0\,\mathfrak{U}(\,\xi_\rho\,,\,\delta)+\,\kappa_2\,\{\,\kappa_0\,\mathfrak{U}(\,\xi_\rho\,,\,\delta)+\,\kappa_1\,\kappa_0\,\mathfrak{U}(\,\xi_\rho\,,\,\delta)\}\,)+\ldots\}]^2\\
&-\vartheta_\rho(1-\vartheta_\rho)\mathfrak{U}^2(\xi_\rho,\mathfrak{J}^{\rho}\xi_\rho)\\
&=(1-\vartheta_\rho)\mathfrak{U}^2(\xi_\rho,\delta)+\vartheta_\rho\mathfrak{U}^2(\xi_\rho,\delta)[\kappa_0+\kappa_1\kappa_0+\kappa_2(\kappa_0+\kappa_1\kappa_0)\\
&+\kappa_3\,(\kappa_0+\kappa_1\,\kappa_0+\kappa_2\,\{\,\kappa_0+\kappa_1\,\kappa_0\,\}\,)\\
&+\ldots+\kappa_{\rho-1}\{\kappa_0+\kappa_1\,\kappa_0+\kappa_2\,\{\,\kappa_0+\kappa_1\,\kappa_0\,\}+\\
&\kappa_3\,(\kappa_0+\kappa_1\,\kappa_0+\kappa_2\,\{\,\kappa_0+\kappa_1\,\kappa_0\,\}\,)+\ldots\}]^2-\vartheta_\rho(1-\vartheta_\rho)\mathfrak{U}^2(\xi_\rho,\mathfrak{J}^{\rho}\xi_\rho)\\
&=(1-\vartheta_\rho)\mathfrak{U}^2(\xi_\rho,\delta)+\vartheta_\rho[\kappa_0(1+\kappa_1)(\,1+\kappa_2\,)(\,1+\kappa_3\,)\ldots(\,1+\kappa_{\rho-1}\,)]^2\\
&\mathfrak{U}^2(\xi_\rho,\delta)-\vartheta_\rho(1-\vartheta_\rho)\mathfrak{U}^2(\xi_\rho,\mathfrak{J}^{\rho}\xi_\rho).
\end{aligned}$$

Since we have $\mathscr{K}=\kappa_0(1+\kappa_1)(\,1+\kappa_2\,)(\,1+\kappa_3\,)\ldots(\,1+\kappa_{\rho-1}\,)<1$.
Therefore,

$$\mathfrak{U}^2(\tau_\rho,\delta)\le\mathfrak{U}^2(\xi_\rho,\delta)-\vartheta_\rho(1-\vartheta_\rho)\mathfrak{U}^2(\xi_\rho,\mathfrak{J}^{\rho}\xi_\rho).$$

Using equations (6) *and* (9), $\lim_{\rho\to\infty}\mathfrak{U}(\mathfrak{J}^{\rho}\xi_\rho,\xi_\rho)=0$.
Also,

$$\begin{aligned}
&\mathfrak{U}(\mathfrak{J}^{\rho+1}\,\xi_\rho,\mathfrak{J}^{\rho+1}\xi_{\rho+1})\le\kappa_0\mathfrak{U}(\xi_{\rho+1},\xi_\rho)+\kappa_1\mathfrak{U}(\mathfrak{J}\,\xi_{\rho+1},\mathfrak{J}\xi_\rho)+\kappa_2\mathfrak{U}(\mathfrak{J}^{2}\xi_{\rho+1},\mathfrak{J}^{2}\xi_\rho)\\
&+\ldots+\kappa_\rho\mathfrak{U}(\mathfrak{J}^{\rho}\xi_{\rho+1},\mathfrak{J}^{\rho}\xi_\rho)\\
&\le\kappa_0\mathfrak{U}(\xi_{\rho+1},\xi_\rho)+\kappa_1\kappa_0\mathfrak{U}(\xi_{\rho+1},\xi_\rho)+\kappa_2\{\kappa_0\mathfrak{U}(\xi_{\rho+1},\xi_\rho)+\kappa_1\kappa_0\mathfrak{U}(\xi_{\rho+1},\xi_\rho)\}\\
&+\ldots+\kappa_\rho\kappa_0\mathfrak{U}(\xi_{\rho+1},\xi_\rho)+\kappa_1\kappa_0+\kappa_2[\{\kappa_0\mathfrak{U}(\xi_{\rho+1},\xi_\rho)+\kappa_1\kappa_0\mathfrak{U}(\xi_{\rho+1},\xi_\rho)\}+\ldots.]\\
&=\kappa_0(1+\kappa_1)(\,1+\kappa_2\,)(\,1+\kappa_3\,)\ldots(\,1+\kappa_\rho\,)\mathfrak{U}(\xi_{\rho+1},\xi_\rho)<\mathfrak{U}(\xi_{\rho+1},\xi_\rho).
\end{aligned}$$

Now,

$$\begin{aligned}
&\mathfrak{U}(\mathfrak{J}\;\;\xi_\rho,\xi_\rho)\\
&\le\mathfrak{U}(\mathfrak{J}\;\;\xi_\rho,\mathfrak{J}^{\rho+1}\xi_\rho)+\mathfrak{U}(\mathfrak{J}^{\rho+1}\;\xi_\rho,\mathfrak{J}^{\rho+1}\xi_{\rho+1})+\mathfrak{U}(\mathfrak{J}^{\rho+1}\xi_{\rho+1},\xi_{\rho+1})+\mathfrak{U}(\xi_\rho,\xi_{\rho+1})\\
&\le\mathfrak{U}(\mathfrak{J}\;\;\xi_\rho,\mathfrak{J}^{\rho+1}\xi_\rho)+\mathfrak{U}(\mathfrak{J}^{\rho+1}\xi_{\rho+1},\xi_{\rho+1})+[1+\kappa_0(1+\kappa_1)(\,1+\kappa_2\,)(\,1+\kappa_3\,)\ldots\\
&(\,1+\kappa_\rho\,)]\mathfrak{U}(\xi_\rho,\xi_{\rho+1})
\end{aligned}$$

which implies that $\lim_{\rho\to\infty}\mathfrak{U}(\mathfrak{J}\xi_\rho,\xi_\rho)=0$.

Now we give the delta-convergence result.

Theorem 10.3 *Let a convex contraction mapping* $\mathfrak{J}$ *of order* ρ *defined from* $\mathscr{H}$ *to* $\mathscr{H}$*, where* $\mathscr{H}$ *is non-empty convex and closed subset of a complete CAT(0) space* $\mathscr{L}$ *and* $\{\xi_\rho\}$ *is an iterative process defined above in* (1) *with* $0<a\le\sigma_\rho,\omega_\rho,\vartheta_\rho\le b<1$. *Then the sequence* $\{\xi_\rho\}$ *delta converges to a fixed point of* $\mathfrak{J}$.

Proof We already have by Lemma 10.1 and 10.2 that $\lim_{\rho\to\infty}\mathfrak{U}(\xi_\rho, \delta)$ exists for $\delta \in \wp(\mathfrak{I})$ and $\lim_{\rho\to\infty}\mathfrak{U}(\mathfrak{I}\xi_\rho, \xi_\rho) = 0$. Thus the sequence $\{\xi_\rho\}$ is bounded.
We now let $\omega_w(\xi_\rho) = \cup A(\{\alpha_\rho\})$, where the union is applied to all sub-sequences $\{\alpha_\rho\}$ of $\{\xi_\rho\}$.
We assert that $\omega_w(\xi_\rho) \subset \wp(\mathfrak{I})$ and is a singleton set, to demonstrate the Δ–convergence of $\{\xi_\rho\}$ to a fixed point of $\mathfrak{I}$. Let $\alpha \in \omega_w(\xi_\rho)$, then $\exists$ a sub-sequence $\{\alpha_\rho\}$ of $\{\xi_\rho\}$ such that $A(\{\alpha_\rho\}) = \{\alpha\}$.
From Lemma 10.5, $\exists$ a sub-sequence $\{\beta_\rho\}$ of $\{\alpha_\rho\}$ such that $\Delta\text{–}\lim_{\rho\to\infty} \beta_\rho = \beta$ for some $\beta \in \mathscr{H}$.
From Lemma 10.11, $\beta \in \wp(\mathfrak{I})$. Also by Lemma 10.1, $\lim_{\rho\to\infty}\mathfrak{U}(\xi_\rho, \beta)$ exists.
Now, we claim that $\alpha = \beta$. Let $\alpha \neq \beta$, then given that asymptotic centers are distinct, so

$$\begin{aligned}\lim\, sup_{\rho\to\infty}\mathfrak{U}(\beta_\rho,\beta) &< \lim\, sup_{\rho\to\infty}\mathfrak{U}(\beta_\rho,\alpha)\\ &= \lim\, sup_{\rho\to\infty}\mathfrak{U}(\alpha_\rho,\alpha)\\ &< \lim\, sup_{\rho\to\infty}\mathfrak{U}(\alpha_\rho,\beta)\\ &= \lim\, sup_{\rho\to\infty}\mathfrak{U}(\xi_\rho,\beta)\\ &= \lim\, sup_{\rho\to\infty}\mathfrak{U}(\beta_\rho,\beta),\end{aligned}$$

which is a contradiction and hence $\alpha = \beta \in \wp(\mathfrak{I})$. Hence $\omega_w(\xi_\rho) \subset \wp(\mathfrak{I})$.
It is sufficient to demonstrate that $\omega_w(\xi_\rho)$ consists of only one point in order to prove that $\{\xi_\rho\}\Delta$–converges to a fixed point of $\mathfrak{I}$. Let $\{\alpha_\rho\}$ be a sub-sequence of $\{\xi_\rho\}$. From Lemma 10.5, $\exists$ a sub-sequence $\{\beta_\rho\}$ of $\{\alpha_\rho\}$ such that $\Delta\text{–}\lim_{\rho\to\infty} \beta_\rho = \beta$ for some $\beta \in \mathscr{H}$. Let $A(\{\alpha_\rho\}) = \{\alpha\}$ and $A(\{\xi_\rho\}) = \{\xi\}$. We have $\alpha = \beta$ and $\beta \in \wp(\mathfrak{I})$. It only remains to prove that $\xi = \beta$. Let $\xi \neq \beta$, then existence of $\lim_{\rho\to\infty}\mathfrak{U}(\xi_\rho, \beta)$ and we know that asymptotic centers are distinct, imply that

$$\begin{aligned}\lim\, sup_{\rho\to\infty}\mathfrak{U}(\beta_\rho,\beta) &< \lim\, sup_{\rho\to\infty}\mathfrak{U}(\beta_\rho,\xi)\\ &= \lim\, sup_{\rho\to\infty}\mathfrak{U}(\xi_\rho,\xi)\\ &< \lim\, sup_{\rho\to\infty}\mathfrak{U}(\xi_\rho,\beta)\\ &= \lim\, sup_{\rho\to\infty}\mathfrak{U}(\beta_\rho,\beta),\end{aligned}$$

which is a contradiction and hence the conclusion follows, that is, $\xi = \beta \in \wp(\mathfrak{I})$. Therefore, $\omega_w(\xi_\rho) = \{\xi\}$.

Theorem 10.4 *Let a convex contraction mapping $\mathfrak{I}$ of order ρ defined from $\mathscr{H}$ to $\mathscr{H}$, where $\mathscr{H}$ is non-empty convex and closed subset of a complete CAT(0) space $\mathscr{L}$ and $\{\xi_\rho\}$ is an iterative process defined above in* (1) *with $0 < a \le \sigma_\rho, \omega_\rho, \vartheta_\rho \le b < 1$, then $\{\xi_\rho\}$ converges strongly to a fixed point of $\mathfrak{I}$ iff $\lim\, inf_{\rho\to\infty}\mathfrak{U}(\xi_\rho, \wp(\mathfrak{I})) = 0$, where $\mathfrak{U}(\xi, \wp(\mathfrak{I})) = inf\{\mathfrak{U}(\xi, \delta) : \delta \in \wp(\mathfrak{I})\}$.*

Proof Necessary condition holds trivially.
Conversely, assume that $lim\; inf_{\rho\to\infty}\, \mathfrak{U}(\xi_\rho, \wp(\mathfrak{I})) = 0$.
From Lemma 10.1, we have $\mathfrak{U}(\xi_{\rho+1}, \delta) = \mathfrak{U}(\xi_\rho, \delta)$ for all $\delta \in \wp(\mathfrak{I})$.
This implies that $\mathfrak{U}(\xi_{\rho+1}, \wp(\mathfrak{I})) = \mathfrak{U}(\xi_\rho, \wp(\mathfrak{I}))$ so that $\lim_{\rho\to\infty} \mathfrak{U}(\xi_\rho, \wp(\mathfrak{I}))$ exists. Thus by hypothesis $\lim_{\rho\to\infty} \mathfrak{U}(\xi_\rho, \wp(\mathfrak{I})) = 0$.
We can now demonstrate that $\{\xi_\rho\}$ is a Cauchy sequence in $\mathscr{H}$. Let $\varepsilon > 0$ be selected at random.
Since $\lim_{\rho\to\infty} \mathfrak{U}(\xi_\rho, \wp(\mathfrak{I})) = 0$, $\exists$ an integer $r_0 \ge 0$ such that
$\mathfrak{U}(\xi_\rho, \wp(\mathfrak{I})) < \varepsilon/4$ for all $\rho \ge r_0$.
In particular, $inf\ \{\mathfrak{U}(\xi_{r_0}, \delta) : \delta \in \wp(\mathfrak{I})\} < \varepsilon/4$. Thus there must exist $\delta^* \in \wp(\mathfrak{I})$ such that $\mathfrak{U}(\xi_{r_0}, \delta^*) < \varepsilon/2$.
Now for all $\rho_1, \rho \ge r_0$, we have

$$\begin{aligned}\mathfrak{U}(\xi_{\rho+\rho_1},\xi_\rho) &= \mathfrak{U}(\xi_{\rho+\rho_1},\delta^*) + \mathfrak{U}(\delta^*,\xi_\rho)\\ &= 2\mathfrak{U}(\xi_{r_0},\ \delta^*)\\ &< 2(\varepsilon/2) = \varepsilon.\end{aligned}$$

Hence $\{\xi_\rho\} \in \mathscr{H} \subseteq \mathscr{L}$, which is a Cauchy sequence; it must converge to a point δ in $\mathscr{H}$, $\lim_{\rho\to\infty} \mathfrak{U}(\xi_\rho, \wp(\mathfrak{I})) = 0$ yields that $\mathfrak{U}(\delta, \wp(\mathfrak{I})) = 0$ and δ must be in $\wp(\mathfrak{I})$ due to the closedness of $\wp(\mathfrak{I})$.

10.4 CONCLUSION

In this article, we have proved results using a general type of mapping, for convex contraction mappings of order ρ and established delta – convergence results and strong convergence results for the same. And every convex contraction of order ρ is Bessaga mapping. Thus our results, of course, hold also for Bessaga mappings in CAT(0) spaces.

REFERENCES

1. M. Abbas, B. S. Thakur and D. Thakur, *Fixed Points of Asymptotically Nonexpansive Mappings in the Intermediate Sense in CAT(0) Spaces.* Commun. Korean Math. Soc., Vol. 28(4), pp. 107-121, 2013.
2. M. Bridson and A. Haefliger, *Metric Spaces of Non-Positive Curvature.* Springer–Verlag, Berlin, Heidelberg, 1999.
3. F. Bruhat and J. Tits, *Groupes Reductifs Sur Un Corps Local. I. Donnees Radicielles Values.* Inst. Hautes Etudes Sci. Publ. Math., Vol. 41, pp. 5-251, 1972.
4. S. Dhompongsa, W. A. Kirk and B. Panyanak, *Nonexpansive Set–valued Mappings in Metric and Banach spaces.* J. Nonlinear Convex Anal., Vol. 8(1), pp. 35-45, 2007.
5. S. Dhompongsa, W. A. Kirk and B. Sims, *Fixed Points of Uniformly Lipschitzian Mappings.* Nonlinear Analysis: Theory, Methods and Applications, Vol. 65(4), pp. 762-772, 2006.
6. S. Dhompongsa and B. Panyanak, *On Δ–convergence theorems in CAT(0) spaces.* Computer and Mathematics with Applications, Vol. 56, pp. 2572-2579, 2008.
7. V. I. Istratescu, *Fixed Point Theorems for Convex Contraction Mappings and Convex Nonexpansive Mappings.* Libertas Mathematica I, 1981.
8. V. I. Istratescu, *Some Fixed Point Theorems for Convex Contraction Mappings and Mappings with Diameters. I.* Ann. Mat. Pura Appl., Vol. 130(4), pp. 89-104, 1982.
9. S. H. Khan and Ritika, *Common Fixed Points of Mean Nonexpansive Mappings in CAT(0) Spaces.* Proceedings of the Jangjeon Mathematical Society, Vol. 23(4), pp. 541-552, 2020.
10. S. H. Khan, Ritika and V. N. Mishra, *A Multistep Iterative Process for Generalized Asymptotically Quasi-Nonexpansive Maps in CAT(0) Spaces.* J. Appl. & Pure Math., Vol. 3(1-2), pp. 75 – 84, 2021.
11. W. A. Kirk, *Geodesic Geometry and Fixed Point Theory.* In Seminar of Mathematical Analysis, (Malaga/Seville, 2002/2003), Colecc. Abierta, Univ. Sevilla Secr. Publ., Seville, Vol. 64, pp. 195-225, 2003.
12. W. A. Kirk, *Geodesic Geometry and Fixed Point Theory II.* In International Conference on Fixed Point Theory and Applications, Yokohama Publ., Yokohama, pp. 113-142, 2004.
13. W. A. Kirk and B. Panyanak, *A Concept of Convergence in Geodesic Spaces.* Nonlinear Analysis: TMA, Vol. 68, pp. 3689-3696, 2008.
14. Ritika, *Common Fixed Points of Generalized Quasi–Contractive Type Operators by a Three Step Iterative Process in CAT(0) Spaces.* International Journal of Mathematics and its Applications, Vol. 6(2-A), pp. 71-77, 2018.
15. Ritika, *Convergence of Three Step Iterative Process for Generalized Nonexpansive Mappings in CAT(0) Spaces.* Aryabhatta Journal of Mathematics and Informatics, Vol. 10(2), pp. 367-372, 2018.
16. Ritika, *Convergence of Three Step Iterative Process for Generalized Asymptotically Quasi–Nonexpansive Mappings in CAT(0) Spaces.* Mathematical Analysis I: Approximation Theory, ICRAPAM 2018, Springer, Vol. 306, pp. 255-261, 2020.
17. Ritika and S.H. Khan, *Convergence of Picard-Mann Hybrid Iterative Process for Generalized Nonexpansive Mappings in CAT(0) Spaces.* Filomat, Vol. 31:11, pp. 3531–3538, 2017. https://doi.org/10.2298/FIL1711531R.
18. Ritika and S. Rathee, *A New Iterative Scheme For Multivalued Mappings in CAT(0) Spaces.* Asian European Journal of Mathematics, Vol. 6(4), pp. 1 – 6, 2013.

19. Ritika and S. Rathee, *Thianwan's Iteration Procedure for Nonexpansive Mappings in CAT(0) Spaces.* Asian European Journal of Mathematics, Vol. 7(1), pp. 1 – 8, 2014.
20. B. S. Thakur, D. Thakur, M. Postolache, *A New Iteration Scheme for Approximating Fixed Points of Nonexpansive Mappings.* Filomat, Vol. 30:10, pp. 2711–2720, 2016. DOI: 10.2298/FIL1610711T.
21. I. Yildirim, Y. Tekmanli and S. H. Khan, *Convex Contractions of Order n in CAT(0) Spaces.* Jordan Journal of Mathematics and Statistics, Vol. 15(1), pp. 123-137, 2022. DOI: https://doi.org/10.47013/15.1.9.

11 Fixed Point Results Comprising with Modular Metric Space and Modular *b*-Metric Space Using Integral Type

Megha S. Pingale
Department of Mathematics, Sandip University, Nashik, India

Renu P. Pathak
Department of Mathematics, Sandip University, Nashik, India

Reeta Bhardwaj
Department of Mathematics, Amity University, Gurugram, Haryana, India

CONTENTS

The objective of the research article to derive common fixed-point theorem for two ρ-compatible functions with integral type in the sense of modular b-metric space, by employing the Δ_2- condition. We generalized the results of Pathak and Bhardwaj [Computer Engineering and Intelligent Systems, 4(9), 2013] in modular b-metric spaces.

11.1 INTRODUCTION

In the field of pure mathematics, the metric fixed point theory offered a wide range of opportunities for innovations to a number of different scholars. Metric is a way of defining the distance that exists between the components of a collection. The metric fixed-point theory has a wide range of applications in the field of real analysis. Numerous uses may be found for it in the fields of engineering and science. The year 1886 marked the beginning of Poincare's work on the fixed point theory. Within the work that he was working on in 1906, Maurice Frechet addressed the concept of metric space.

DOI: 10.1201/9781032634142-11

Banach proved significant contraction mapping principle for the research work in the year 1922. If $(\mathfrak{S}, \partial)$ be a metric space. A self-mapping defined on $\mathscr{S}$ is a contraction mapping if

$$\partial(\mathrm{T}(\sigma), \mathrm{T}(\rho)) \leq \mathrm{k}\partial(\sigma, \rho)$$

for all, $\sigma, \rho \in \mathfrak{S}$ where $0 \leq \mathrm{k} < 1$.

Many researchers functioned on the same path by varying contraction principle with various metric spaces by applying unlike types of mappings. Some other known results on metric spaces and its generalization can be found in [17, 18, 27, 28].

Encouraged by work of Nakano1950 [22], V. Chistyakov [5, 6] in the year 2006 introduced modular metric on a set, The idea of modular metric spaces was introduced in [5, 6]. Let us see modular metric space introduced by Nakano [22] on vector space and a modular function space was studied by Musielak, Orlicz and other authors [19, 20, 21]. Also L. Maligranda studied Orlicz spaces and its interpolation with some new ideas [16].

In the research paper [8], Gholidahneh et al introduced the notion of a modular p-metric space (an extended modular b-metric space) by applying Meir-Keeler contractions in this new space. As per the author the class of modular p-metric spaces is considerably larger than the class of modular b-metric spaces. In [20] Mongkolkeha et. al. modified their results to gain the expected new results. Mohamed A Khamsi [14, 15] presented the research work that includes a series of fixed-point results for pointwise contractions and asymptotic pointwise contractions in reference of modular functions spaces. Author [2] explore the fixed-point property in the modular metric spaces. In the year 2015, M. R. Alfuraidan presented the generalization of the Banach contraction principle on a modular metric space endowed with a graph [3]. Here author discussed the existence of fixed points for single Lipschitzian mappings defined on some subsets of modular metric spaces $\mathfrak{S}$ endowed with a graph G. In the year 2018 Ege et al [7] defined modular b-metric space and proved Banach contraction principle with an application of this theory to the solution of linear equations. Ege et al [7] applied v-contraction to get their result.

Branciari introduced the idea of integral type contractions in metric spaces and letter was same utilized by various researchers to extended several results for such type of contractions in relevant spaces. Some of them are [4, 9, 12].

Let v be a modular b-metric on a set $\mathfrak{S}$ and if T is an arbitrary self-map on X_v^*. For all $x, y \in \mathfrak{S}_v^*, \lambda > 0$ and $0 < k < 1$, such that

$$v_\lambda(\mathrm{T}(\sigma), \mathrm{T}(\rho)) \leq k v_\lambda(\sigma, \rho)$$

then the map T is said to be a v-contraction.

Haokip et. al. verified fixed-point theorems for generalized Kannan type mappings in b metric spaces [3], while in [24], author verified fixed point results for generalized F-contractions with modular b-metric spaces and established results in fuzzy b-metric spaces also.

Abdou [1] studied the existence of fixed points for contractive and non-expansive Kannan maps in the modular metric spaces. It represents successive approximations of fixed points which converge to the fixed points in the modular sense. Turkoglu et. al. [26] introduces a new concept of generalized modular metric space along-with some examples. Authors in [23] introduce the concept of Δ_3 type condition in modular G-metric spaces along with proof of fixed point of some generalized contractible operators in modular G-metric spaces.

11.2 PRELIMINARIES

Definition 11.1 *[8] A function $\xi : (0,\infty) \times X \times X \to [0,+\infty]$ is called a modular metric on X if the following axioms hold:*

1. $x = y$ *if and only if* $\xi_\lambda(x,y) = 0$ *for all* $\lambda > 0$,
2. $\xi_\lambda(x,y) = \xi_\lambda(y,x)$ *for all* $\lambda > 0$ *and* $x,y \in X$,
3. $\xi_{\lambda+\mu}(x,y) \leq \xi_\lambda(x,z) + \xi_\mu(z,y)$ *for all* $\lambda,\mu > 0$ *and* $x,y,z \in X$

A modular metric ξ on X is called regular if the following weaker version of (i) is satisfied:

$$x = y \text{ if and only if } \xi_\lambda(x,y) = 0$$

for some $\lambda > 0$. The alternative definition of modular metric space is given below:

Definition 11.2 *[25] Let ξ be an arbitrary vector space over* $K = (R$ *or* $C)$. *A functional* $F : \xi \to [0,\infty]$ *is called modular if*

1. $F(x) = 0$ *iff* $x = 0$,
2. $F(\alpha x) = F(x)$ *for* $\alpha \in K$ *with* $|\alpha| = 1$ *for all* $x \in X$,
3. $F(\alpha x + \beta y) \leq F(x) + F(y)$ *if* $\alpha,\beta \geq 0, \alpha+\beta = 1$, *for all* $x,y \in X$

If (iii) is replaced by

4. $F(\alpha x + \beta y) \leq \alpha F(x) + \beta F(y)$ *if* $\alpha,\beta \geq 0, \alpha+\beta = 1$, *for all* $x,y \in X$

Then the modular F is called convex modular.

A modular ξ_ρ defines a corresponding modular space; i. e. the space ξ_ρ given by:

$$\xi_\rho = \{x \in X; F(\alpha x) \to 0 \text{ as } \alpha \to 0\},$$

Turkoglu et. al. [26] defined the concept of generalized modular metric spaces.

Definition 11.3 *[26] Let X be an abstract set. A function* $D : (0,\infty) \times X \times X \to [0,\infty]$ *is said to be a regular generalized modular metric (GMM) on X if it satisfies the following three axioms:*

1. *If* $D_\lambda(x,y) = 0$ *for some* $\lambda > 0$, *then* $x = y \ \forall x,y \in X$,
2. $D_\lambda(x,y) = D_\lambda(y,x) \ \forall \lambda > 0$ *and* $x,y \in X$,
3. $\exists C > 0$: *if* $(x,y) \in X \times X, \{x_n\} \subset X$ *with* $\lim_{n\to\infty} D_\lambda(x_n,x) = 0$ *for some* $\lambda > 0$,

then

$$D_\lambda(x,y) \leq \limsup_{n\to\infty} D_\lambda(x_n,y),$$

The pair (X,D) *is said to be a generalized modular metric space (GMMS).*

Definition 11.4 *[25] Let ξ_ρ be a modular space. A sequence $(x_n)_{n\in N}$ in ξ_ρ is said to be;*

1. $\rho-$ *convergent to* x *if* $\rho(x_n - x) \to 0$ *as* $n \to \infty$,
2. $\rho-$ *Cauchy if* $\rho(x_n - x_m) \to 0$ *as* $n \to \infty$,

3. ξ_ρ is ρ-complete if every $\rho-$ Cauchy sequence is ρ-convergent,

4. A subset $B \subset \xi_\rho$ is said to be ρ-closed if for any sequence $(x_n)_{n\in N} \subset B$ and $x_n \to x$, We have $x \in B$,

5. A subset $B \subset \xi_\rho$ is said to be ρ-bounded if $\delta_\rho(B) = \sup \rho(x-y) < \infty$ for all, $x,y \in B$, where $\delta_\rho(B)$ is the ρ-diameter of B.,

6. Fatou property for ρ defined as if $\rho(x-y) \leq \operatorname{lin}\inf \rho(x_n - y_n)$ whenever $x_n \to x$ and $y_n \to y$ as $n \to \infty$.

7. Δ_2 - condition for ρ defined as if $\rho(2x_n) \to 0$ whenever $\rho(x_n) \to 0$ as $n \to \infty$

Definition 11.5 *[25] Let ξ_p be a modular space, where ρ satisfy the Δ_2-condition. Two self-mappings g and R of ξ_p are called ρ-compatible if $(gRx_n - Rgx_n) \to 0$ whenever $(x_n)_{n\in N}$ is a sequence in ξ_p such that $Rx_n \to z$ and $gx_n \to z$ for some $z \in \xi_p$*

Ege and Alaca [7] introduced the notion of modular b-metric space.

Definition 11.6 *[7] Let X be a non-empty set and $s \geq 1$ a real number. A mapping $\zeta : (0,\infty) \times X \times X \to [0,\infty]$ is called a modular metric on X, if the following axioms hold for all, $x,y,z \in X$:*

1. *$\zeta_\lambda(x,y) = 0$ for all $\lambda > 0$ if and only if $x = y$,*
2. *$\zeta_\lambda(x,y) = \zeta_\lambda(y,x)$ for all $\lambda > 0$,*
3. *$\zeta_{\lambda+\mu}(x,y) \leq s\left[\zeta_\lambda(x,z) + \zeta_\mu(z,y)\right]$ for all $\lambda, \mu > 0$*

Then (X,ζ) is called a modular b -metric space.

Definition 11.7 *[25] Let (X,d) be a metric space. A self-mapping $T : X \to X$ is called quasi contraction if and only if for all $x,y \in X$ there exists $h < 1$ and*

$$d(Tx,Ty) \leq h\max\{d(x,y),d(x,Tx),d(y,Ty),d(x,Ty),d(y,Tx)\}$$

Example 11.1 *[7] Consider the space*

$$l_p = \left\{(x_n) \subset R : \sum_{n=1}^{\infty} |x_n|^p < \infty\right\}, 0 < p < 1, \lambda \in (0,\infty)$$

and

$$v_\lambda(x,y) = \frac{d(x,y)}{\lambda},$$

such that

$$d(x,y) = \left(\sum_{n=1}^{\infty} |x_n - y_n|^p\right)^{\frac{1}{p}}, x = x_n, y = y_n \in l_p$$

It could be easily seen that (x,v) is a modular b-metric space.
Indeed, it is obvious that

$$d(x,z) \leq 2^{\frac{1}{p}}[d(x,y) + d(y,z)]$$

is a b-metric.

$$\begin{aligned} v_{\lambda+\mu}(x,z) &= \frac{d(x,z)}{\lambda+\mu} \\ &\leq 2^{\frac{1}{p}}\left[\frac{d(x,y)}{\lambda+\mu}+\frac{d(y,z)}{\lambda+\mu}\right] \\ &\leq 2^{\frac{1}{p}}\left[\frac{d(x,y)}{\lambda}+\frac{d(y,z)}{\mu}\right] \\ &= 2^{\frac{1}{p}}\left[v_{\lambda}(x,y)+v_{\mu}(y,z)\right] \end{aligned}$$

11.3 MAIN RESULTS

The primary result of our paper is the following theorem.

Theorem 11.1 *Let ξ_p be a modular b-metric space with coefficient $s \geq 1$ and $g, R : \xi_p \to \xi_p$ with ρ-compatible. Then the function $R(\xi_p) \subseteq g(\xi_p)$ and suppose $a, b, c \in R^+, a > c$ with* g *and* R *as two compatible mappings, and such that*

$$\int_0^\rho [a(R_x - R_y)] F(t)dt \leq u\int_0^\rho \left[c\frac{(x,g_x)-(x,g_y)}{g_x g_y}\right] F(t)dt + v\int_0^\rho \left[\frac{\frac{(x,g_x)-(x,g_y)}{g_x g_y} + (x_{n+1}, g_{x_{n+1}}) - (x_n, R_{x_n})}{1+\frac{(x,g_x)-(x,g_y)}{g_x g_y}(x_{n+1}, g_{x_{n+1}}) - (x_n, R_{x_n})}\right] F(t)dt \tag{11.1}$$

for some $b \in (0,1)$ where $F : R^+ \to R^+$ is a mapping which is Lebesgue integrable and which is summable and always positive as $\in > 0$ such that

$$\int_0^\varepsilon F(t)dt > 0 \tag{11.2}$$

whereas the function g and R are continuous and non-decreasing, then there exists a unique common fixed point of maps g and R.

Proof *As $\lambda \in R^+, \frac{a}{c}+\frac{1}{\lambda}=1$. Let y_0 be any arbitrary point of ξ_p and generate inductively sequence $\{Rx_n\}, n \in N$, as $R(\xi_p) \subseteq g(\xi_p)$.*
For each integer $n \geq 1$, Eq. (11.1) show that

$$\int_0^\rho [a(R_{x_{n+1}} - R_{x_n})] F(t)dt \leq u\int_0^\rho \left[c\frac{(x_{n+1},g_{x_{n+1}})-(x_{n+1},g_{x_n})}{g_{x_{n+1}} g_{x_n}}\right] F(t)dt + v\int_0^\rho \left[c\frac{\left\{\frac{(x_{n+1},g_{x_{n+1}})-(x_{n+1},g_{x_n})}{g_{x_{n+1}} g_{x_n}}\right\} + \{(x_{n+1}, g_{x_{n+1}}) - (x_n, R_{x_n})\}}{1+\left\{\frac{(x_{n+1},g_{x_{n+1}})-(x_{n+1},g_{x_n})}{g_{x_{n+1}} g_{x_n}}\right\}\{(x_{n+1}, g_{x_{n+1}}) - (x_n, R_{x_n})\}}\right] F(t)dt.$$

Since $T_{x_n} = T_{x_{n-1}}$ then,

$$\int_0^\rho [a(R_{x_{n+1}} - R_{x_n})] F(t)dt \leq u\int_0^\rho \left[c\frac{(x_n,R_{x_n})-(x_n,R_{x_{n-1}})}{R_{x_n} R_{x_{n-1}}}\right] F(t)dt + v\int_0^\rho \left[c\frac{\left\{\frac{(x_n,R_{x_n})-(x_n,R_{x_{n-1}})}{R_{x_n} R_{x_{n-1}}}\right\}+0}{1+0}\right] F(t)dt,$$

$$\int_0^{\rho}\left[a\left(R_{x_{n+1}}-R_{x_n}\right)\right]F(t)dt \le u\int_0^{\rho}\left[c\left\{\frac{(x_n,R_{x_n})-(x_n,R_{x_{n-1}})}{R_{x_n}R_{x_{n-1}}}\right\}\right]F(t)dt$$
$$+v\int_0^{\rho}\left[c\left\{\frac{(x_n,R_{x_n})-(x_n,R_{x_{n-1}})}{R_{x_n}R_{x_{n-1}}}\right\}\right]F(t)dt,$$
$$\le (u+v)\int_0^{\rho}\left[c\left\{\frac{(x_n,R_{x_n})-(x_n,R_{x_{n-1}})}{R_{x_n}R_{x_{n-1}}}\right\}\right]F(t)dt,$$

Let $u+v=b$

$$\int_0^{\rho}\left[a\left(R_{x_{n+1}}-R_{x_n}\right)\right]F(t)dt \le b\int_0^{\rho}\left[c\left\{\frac{(x_n,R_{x_n})-(x_n,R_{x_{n-1}})}{R_{x_n}R_{x_{n-1}}}\right\}\right]F(t)dt,$$

Applying induction we get,

$$\int_0^{\rho}\left[a\left(R_{x_{n+1}}-R_{x_n}\right)\right]F(t)dt \le b^n\int_0^{\rho}\left[c\frac{(x,R_x)-(x,x)}{R_x x}\right]F(t)dt$$

Taking limit $n\to\infty$ *yields,*

$$\lim_{n\to\infty}\int_0^{\rho}\left[a\left(R_{x_{n+1}}-R_{x_n}\right)\right]F(t)dt \le 0 \tag{11.3}$$

Now we can prove that $\{R_{x_n}\},\eta_0\in N$ *is a* ρ*-Cauchy Sequence. If not then for given* $\in>0$. *Let the sequences* $\{\{\eta_{0(\Delta)}\},\{\mu_{0(\Delta)}\}\}$ *with* $\eta_{0(\Delta)}>\mu_{0(\Delta)}\ge\Delta$, *such that*

$$d_\Delta=\rho\left[c\left(R_{\eta_{0(\Delta)-1}}-R_{\mu_{0(\Delta)}}\right)\right]\ge\varepsilon,\forall\Delta=1,2,3, \tag{11.4}$$

We assume that

$$\rho\left[c\left(R_{\eta_{0(\Delta)-1}}-R_{\mu_{0(\Delta)}}\right)\right]<\varepsilon \tag{11.5}$$

In order to show this, suppose $\eta_{0(\Delta)}$ *is the smallest number exceeding* $\mu_{0(\Delta)}$ *for which (11.4) holds and* $\Sigma_\Delta=\left\{\eta_0\in N:\exists\mu_{0(\Delta)}\in \mathrm{N}:\rho\left[c\left(R_{x_{\eta_{0(\Delta)-1}}}-R_{x_{\mu_{0(\Delta)}}}\right]\ge\varepsilon\right.\right.$, *and* $\eta_{0(\Delta)}>\mu_{0(\Delta)}\ge\Delta$, *Evidently* $\Sigma_\Delta\ne\emptyset$ *and since by the principle of ordering* $\Sigma_\Delta\subseteq N$ *and the elements of* Σ_Δ *is denoted by* $\eta_{0(\Delta)}$ *as a sequence function. Obviously, equation (11.5) holds.*
Further, on solving we get

$$\int_0^{\rho}\left[a\left(R_{x_{\mu_{0(\Delta)}}}-R_{x_{\eta_{0(\Delta)}}}\right)\right]F(t)dt$$
$$\le b\int_0^{\rho}\left[c\left\{\frac{\left(x_{\mu_{0(\Delta)-1}},R_{x_{\mu_{0(\Delta)-1}}}\right)-\left(x_{\mu_{0(\Delta)-1}},R_{x_{\eta_{0(\Delta)-1}}}\right)}{R_{x_{\mu_{0(\Delta)-1}}}R_{x_{\eta_{0(\Delta)-1}}}}\right\}\right]F(t)dt,$$

Condition by (11.3) we have,

$$\rho\left[c\left\{\frac{\left(x_{\mu_0(\Delta)-1},R_{x_{\mu_0(\Delta)-1}}\right)-\left(x_{\mu_{0(\Delta)-1}},R_{x_{\eta_0(\Delta)-1}}\right)}{R_{x_{\mu_0(\Delta)-1}},R_{x_{\eta_0(\Delta)-1}}}\right\}\right]$$

$$=\rho\left[c\left\{\frac{\left(x_{\mu_0(\Delta)-1},R_{x_{\mu_0(\Delta)-1}}\right)-\left(x_{\mu_0(\Delta)},R_{x_{\mu_0(\Delta)}}\right)+\left(x_{\mu_0(\Delta)},R_{x_{\mu_0(\Delta)}}\right)-\left(x_{\mu_0(\Delta)-1},R_{x_{\eta_0(\Delta)-1}}\right)}{R_{x_{\mu_0(\Delta)-1}},R_{x_{\eta_0(\Delta)-1}}}\right\}\right]$$

$$=\rho\left[\frac{\frac{\propto c}{\propto}\left\{\left(x_{\mu_0(\Delta)-1},R_{x_{\mu_0(\Delta)-1}}\right)-\left(x_{\mu_0(\Delta)},R_{x_{\mu_0(\Delta)}}\right)\right\}}{R_{x_{\mu_0(\Delta)-1}},R_{x_{\eta_0(\Delta)-1}}}\right]$$

$$+\rho\left[\frac{\frac{c\beta}{\beta}\left\{\left(x_{\mu_0(\Delta)},R_{x_{\mu_0(\Delta)}}\right)-\left(x_{\mu_0(\Delta)-1},R_{x_{\eta_0(\Delta)-1}}\right)\right\}}{R_{x_{\mu_0(\Delta)-1}},R_{x_{\eta_0(\Delta)-1}}}\right]$$

$$=\rho\left[\frac{\propto c\left\{\left(x_{\mu_0(\Delta)-1},R_{x_{\mu_0(\Delta)-1}}\right)-\left(x_{\mu_0(\Delta)},R_{x_{\mu_0(\Delta)}}\right)\right\}}{R_{x_{\mu_0(\Delta)-1}},R_{x_{\eta_0(\Delta)-1}}}\right]$$

$$+\rho\left[\frac{\beta\left\{\left(x_{\mu_0(\Delta)},R_{x_{\mu_0(\Delta)}}\right)-\left(x_{\mu_{0(\Delta)-1}},R_{x_{\eta_0(\Delta)-1}}\right)\right\}}{R_{x_{\mu_0(\Delta)-1}},R_{x_{\eta_0(\Delta)-1}}}\right]$$

Using Δ_2- condition and (11.3) then,

$$\lim_{\Delta\to 0}\rho\left[\frac{\propto_c\left\{\left(x_{\mu_{0(\Delta)-1}},,R_{x_{\mu_0(\Delta)-1}}\right)-\left(x_{\mu_{0(\Delta)}},,R_{x_{\mu_0(\Delta)}}\right)\right\}}{R_{x_{\mu_0(\Delta)-1}},R_{x_{\eta_0(\Delta)-1}}}\right]=0$$

Therefore

$$\lim_{\Delta}\int_0^{\rho}\left[c\left\{\frac{\left(x_{\mu_{0(\Delta)-1}},,R_{x_{\mu_{0(\Delta)-1}}}\right)-\left(x_{\mu_{0(\Delta)-1}},R_{x_{\eta_0(\Delta)-1}}\right)}{R_{x_{\mu_{0(\Delta)-1}}}\ R_{x_{\eta_0(\Delta)-1}}}\right\}\right]F(t)dt\leq\int_0^{\varepsilon}F(t)dt,$$

Also, we have shown that the result (11.4) that it gives a contradiction. Therefore, it is a $\rho-$ Cauchy sequence, since ξ_p is ρ - complete then there exist a point $z_0\in\xi_p$ such that the sequence $\rho\left[a\left(Rx_n-z_0\right)\to 0\right.$ as $n\to\infty$

If the mapping R is continuous then these sequences converge i. e. $R^2x_n\to Rz_0$ and $Rg_{x_n}\to Rz_0$, Hence now we prove that z_0 is a fixed point of R we have taking as $n\to\infty$

$$\int_0^{\rho}\left[a\left(Rz_0-z_0\right)\right]F(t)dt\leq b\int_0^{\rho}\left[c\frac{\left(Rz_0-z_0\right)}{Rz_0z_0}\right]F(t)dt$$

this gives

$$\int_0^{\rho}\left[a\left(Rz_0-z_0\right)\right]F(t)dt\leq 0,$$

Since by (11.2) $\rho\left[a\left(Rz_0-z_0\right)\right]\to 0$ and $Rz_0=z_0$.

Also $R(\xi_\rho) \subseteq g(\xi_\rho)$. Then there exists a point m *such that $z_0 = Rz_0 = g_m$. Then the inequality at $n \to \infty$ gives*

$$\int_0^\rho [a(Rz_0 - Rm)]F(t)dt \leq b\int_0^\rho \left[\frac{(z_0, Rz_0) - (z_0, gm)}{Rz_0 gm}\right] F(t)dt,$$

$$\int_0^\rho [a(z_0 - Rm)]F(t)dt \leq \mathrm{b}\int_0^\rho \left[c\frac{(z_0, z_0) - (z_0, gm)}{z_0 gm}\right] F(t)dt,$$

$$\int_0^\rho [a(z_0 - R_m)]F(t)dt \leq b\int_0^\rho [c(z_0 - z_0)]F(t)dt,$$

Therefore, as a result is $z_0 = R_m = g_m$ and also $gz_0 = gR_m = Rg_m = Rz_0 = z_0$. Our assumption that z_0 and z_2 are two common fixed points of R *and* g*, then we get $\rho[a(z_0 - z_2) = 0$*
Hence, we have $z_0 = z_2$ is a common fixed point of function R *and* g.

Theorem 11.2 *Let ξ_ρ be a modular b-metric space with coefficient $s \geq 1$ and ζ_ρ is a subset of ξ_ρ that is $\rho-$ closed and $\rho-$ bounded and $g, R : \zeta_\rho \to \zeta_\rho$ with ρ-compatible then the function $R(\xi_\rho) \subseteq g(\xi_\rho)$ and*

$$\int_0^\rho [a(R_x - R_y)]F(t)dt \leq u\int_0^\rho \left[a\left\{\frac{(x, g_x) - (x, g_y)}{g_x g_y}\right\}\right] F(t)dt$$
$$+v\int_0^\rho \left[a\left\{\frac{\left\{\frac{(x,g_x)-(x,g_y)}{g_x g_y}\right\} + \left\{\left(x_{\eta_0+1}, g_{x_{\eta_0+1}}\right) - \left(x_{\eta_0}, g_{x_{\eta_0}}\right)\right\}}{\left\{\frac{(x,g_x)-(x,g_y)}{g_x g_y}\right\}\left(x_{\eta_0+1}, g_{x_{\eta_0+1}}\right) - \left(x_{\eta_0}, g_{x_{\eta_0}}\right)\right\}}\right\}\right] F(t)dt, \tag{11.6}$$

$\forall x, y \in \zeta_\rho$. Also $a, b \in R^+$ with $b \in (0,1)$ and the mapping for the function $F : R^+ \to R^+$ is a Lebesgue integrable mapping which is addible and always positive

$$\int_0^\varepsilon F(t)dt > 0, \quad \forall \in > 0$$

Further, if one of function g and R is continuous then $\exists$ a unique fixed point of g and R.

Proof *Let $x \in \zeta_\rho$ and $\mu_0, \eta_0 \in N$ then from (11.6) on continuing iteration n-times, we get*

$$\int_0^\rho \left[a\left(R_{x\eta_0+\mu_0} - R_{x_{\mu_0}}\right)\right] F(t)dt \leq b^n \int_0^\rho [a(R_{x\eta_0} - x)]F(t)dt$$

$$\int_0^\rho \left[a\left(R_{x\eta_0+\mu_0} - R_{x_{\mu_0}}\right)\right] F(t)dt \leq b^n \int_0^{\varphi_\rho} \zeta_\rho F(t)dt$$

Since ζ_ρ is ρ- bounded, then

$$\lim_{\eta_0,\mu_0 \to \infty} \int_0^\rho \left[a\left(R_{x_{\eta_0}+\mu_0} - R_{x_{\mu_0}}\right)\right] F(t)dt \leq 0,$$

Which implies that

$$\lim_{\eta_0,\mu_0 \to \infty} \rho\left[a\left(R_{x_{\eta_0+\mu_0}} - R_{x_{\mu_0}}\right)\right] \to 0,$$

Therefore, by the Δ_2 - condition $\left\{R_{x_{\eta_0}}\right\}_{\eta_0 \in N}$ is a $\rho-$ Cauchy sequence and also by our assumption ξ_ρ, is $\rho-$ complete and ζ_0 is $\rho-$ closed then there exists a point $z_0 \in B$, such that

$$\lim_{n \to \infty} \rho\left[a\left(R_{x_{\eta_0}} - z_0\right)\right] \to 0,$$

If a continuous mapping R*, we have* $R^2_{x_{\eta_0}} \to Rz_0$ *and* $Rg_{x_{\eta_0}} \to Rz_0$*, then* $\rho\left(gR_{x_{\eta_0}} - Rg_{x_{\eta_0}}\right) \to 0$ *Therefore by* ρ*-compatibility* $gR_{x_{\eta_0}} \to Rz_0$*. Hence we show that* z_0 *is a fixed point of* R *such as* $Rz_0 = gz_0 = z_0$ *Finally, let* z_0 *and* v_0 *be two arbitrary common fixed points of* R *and* g*, then*

$$\int_0^\rho [a(z_0 - v_0)] F(t)dt \leq u \int_0^\rho \left[a \left\{ \frac{(z_0, Rz_0) - (v_0, Rv_0)}{Rz_0 Rv_0} \right\} \right] F(t)dt,$$

$$\int_0^\rho [a(z_0 - v_0)] F(t)dt \leq u \int_0^\rho \left[a \left\{ \frac{(z_0, z_0) - (v_0, v_0)}{z_0 v_0} \right\} \right] F(t)dt$$

which gives as $\rho[a(z_0, z_0) - (v_0, v_0)] \to 0$*, i.e., we say* $\rho[a(z_0 - v_0)] \to 0$ *and hence then we have* $z_0 = v_0$.

Some consequence of our finding is as follows:

Corollary 11.1 *Let* ξ_p *be a modular b-metric space with coefficient* $s \geq 1$ *and* $g : \xi_p \to \xi_p$ *with* $\rho-$ *compatible. Suppose* $a, b, c \in R^+, a > c$ *with g as a continuous mapping.*

$$\int_0^\rho [a(g_x - g_y)] F(t)dt \leq u \int_0^\rho \left[c \frac{(x, g_x) - (x, g_y)}{g_x g_y} \right] F(t)dt$$
$$+ v \int_0^\rho \left[\frac{\frac{(x,g_x)-(x,g_y)}{g_x g_y} + (x_{n+1}, g_{x_{n+1}}) - (x_n, g_{x_n})}{1 + \left\{ \frac{(x,g_x)-(x,g_y)}{g_x g_y} \right\} (x_{n+1}, g_{x_{n+1}}) - (x_n, g_{x_n})} \right] F(t)dt$$

for some $b \in (0,1)$ *where* $F : R^+ \to R^+$ *is a mapping which is Lebesgue integrable and which is summable and always positive as* $\in > 0$ *such that* $\int_0^\varepsilon F(t)dt > 0$*. Then* $\exists$ *a unique fixed point of g.*

Corollary 11.2 *Let* ξ_ρ *be a modular b-metric space with coefficient* $s \geq 1$ *and* ζ_ρ *is a subset of* ξ_ρ *that is* $\rho-$ *closed and* $\rho-$ *bounded and* $g : \zeta_\rho \to \zeta_\rho$ *with* ρ*-compatible then the function and*

$$\int_0^\rho [a(g_x - g_y)] F(t)dt \leq u \int_0^\rho \left[a \left\{ \frac{(x, g_x) - (x, g_y)}{g_x g_y} \right\} \right] F(t)dt$$
$$+ v \int_0^\rho \left[a \left\{ \frac{\left\{ \frac{(x,g_x)-(x,g_y)}{g_x g_y} \right\} + \left\{ \left(x_{\eta_0+1}, g_{x_{\eta_0+1}} \right) - \left(x_{\eta_0}, g_{x_{\eta_0}} \right) \right\}}{\left\{ \frac{(x,g_x)-(x,g_y)}{g_x g_y} \right\} \left(x_{\eta_0+1}, g_{x_{\eta_0+1}} \right) - \left(x_{\eta_0}, g_{x_{\eta_0}} \right) \right\}} \right\} \right] F(t)dt,$$

$\forall x, y \in \zeta_\rho$ *Also* $a, b \in R^+$ *with* $b \in (0,1)$ *and the mapping for the function* $F : R^+ \to R^+$ *is a Lebesgue integrable mapping which is addible and always positive* $\int_0^\varepsilon F(t)dt > 0, \forall \in > 0$ *Further, if the function g is continuous then* $\exists$ *a unique fixed point of g.*

REFERENCES

1. Abdou, A. A. N. (2020). Fixed points of Kannan maps in modular metric spaces. AIMS Mathematics, 5(6), 6395–6403. DOI: 10.3934/math.2020411
2. Abdou, A. A. N., & Khamsi, M. A. (2013). Fixed point results of pointwise contractions in modular metric spaces. Fixed Point Theory and Applications, 2013, Article Id: 163.
3. Alfuraidan, M. R. (2015). The contraction principle for mappings on a modular metric space with a graph. Fixed Point Theory and Applications, 2015, Article Id: 46, DOI 10.1186/s13663-015-0296-3
4. Bhardwaj, R. (2012). Some common fixed-point theorem in metric space using integral type mappings. IOSR Journal of Engineering, 2, 187–190.
5. Chistyakov, V.V. (2010). Modular metric spaces, I: Basic concepts. Nonlinear Anal., 72, 1–14.

6. Chistyakov, V.V. (2010). Modular metric spaces, II: Application to superposition operators. Nonlinear Anal., 72, 15–30.
7. Ege, M. E., Alaca, C. (2018). Some results for modular b-metric spaces and an application to system of linear equations. Azerb. J. Math., 8, 3–14.
8. Gholidahneh, A. Sedghi, S. Ege,O. Mitrovic, Z. D. and De la Sen, M. (2021). The Meir-Keeler type contractions in extended modular b-metric spaces with an application. AIMS Mathematics, 6(2), 1781–1799. DOI: 10.3934/math.2021107
9. Gupta, V. Mani, N., Ansari,A. H. (2018). Generalized integral type contraction and common fixed-point theorems using an auxiliary function. Advances in Mathematical Sciences and Applications, 27(2), 263–275.
10. Gupta, V. Mani, N. (2013). Existence and uniqueness of fixed point for contractive mapping of integral type. Int. J. Computing Science and Mathematics, 4(1), 72–83,
11. Gupta, V. Mani, N. (2013). Common Fixed Point for Two Self-maps Satisfying a Generalized $\Psi \int_{\phi}$ - Weakly Contractive Condition of Integral Type. International Journal of Nonlinear Science, 16(1), 64–71.
12. Gupta, V. Mani, N. (2011). A Common Fixed-Point Theorem for Two Weakly Compatible Mappings Satisfying a New Contractive Condition of Integral Type. Mathematical Theory and Modeling, 1(1), 146–148.
13. Haokip, N. Goswami, N. (2019). Some fixed-point theorems for generalized Kannan type mappings in b metric spaces. Proyecciones (Antofagasta online), 38(4), 763–783.
14. Khamsi, M. A., Kozlowski,W. K. (2010). On asymptotic point wise contractions in modular function spaces. Nonlinear Anal., 73, 2957–2967.
15. Khamsi, M. A., Kozlowski,W. K. (2011). On asymptotic pointwise nonexpansive mappings in modular function spaces. J. Math. Anal. Appl., 380, 697–708.
16. Maligranda, L. (1989). Orlicz spaces and interpolation", Semin. in Math., 5, 1–14.
17. Mani, N. (2018). Generalized C_{β}^{ψ} - rational contraction and fixed point theorem with application to second order deferential equation. Mathematica Moravica, 22(1), 43–54.
18. Gupta, V., Jungck, G., Mani, N. (2020). Some novel fixed point theorems in partially ordered metric spaces. AIMS Mathematics, 5(5), 4444–4452.
19. Mongkolkeha, C., Sintunavarat, W., Kumam,P. (2011). Fixed point theorems for contraction mappings in modular metric spaces. Fixed Point Theory Appl.,2011, Article Id:93, 9 pages.
20. Musielak, J. Orlicz,W.(1959). On modular spaces. Studia Math., 18, 49–65.
21. Musielak, J. Orlicz,W.(1959). Some remarks on modular spaces. Bull. Acad. Pol. Sci. Ser. Sci. Math., Astron. Phys., 7, 661–668.
22. Nakano, H. (1950). Modulared semi-ordered linear spaces. Tokyo: Maruzen.
23. Okeke,G.A. Francis,D., De la Sen, M., Abbas, M. (2021). Fixed point theorems in modular G- metric spaces. Journal of Inequalities and Applications, 2021, Article Id:163. https://doi.org/10.1186/s13660-021-02695-8
24. Parvaneh,V., Hussain, N., Khorshidi,M., Mlaiki, N., Aydi,H., Fixed Point Results for Generalized F-Contractions in Modular b-Metric Spaces with Applications. Mathematics, 7, Article Id: 887. doi:10.3390/math7100887
25. Pathak,R.P., Bhardwaj,R., Common Fixed-point theorems for contractive maps of integral type in modular metric spaces. Computer Engineering and Intelligent Systems, 4(9), 131–140.
26. Turkoglu, D. Manav, N. (2018). Fixed point theorems in a new type of modular metric spaces. Fixed Point Theory and Applications 2018, Article Id:25. https://doi.org/10.1186/s13663-018-0650-3
27. Mani, N., Beniwal,S., Shukla, S., Pingale, M. (2023). Fixed Point Theory in Extended Parametric Sb-Metric Spaces and Its Applications. Symmetry, 15 — Article ID 2136.
28. Mani, N., Pingale,M., Shukla, S., Pathak, R. (2023). Fixed point theorems in fuzzy b-metric spaces using two different t-norms. Adv. Fixed Point Theory, 13, Article ID 29.

12 Properties of $(m,n)^*$-Paranormal Operators: Composition, Multiplicative, and Weighted Forms

Baljinder Kour
Department of Mathematics, Akal University, Talwandi Sabo, Bathinda, Punjab, Inida

Ashish Arora
Department of Mathematics, IKG Punjab Technical University, Kapurthala, Punjab, India

CONTENTS

In this paper, we prove some characterizations for the class of $(m,n)^*$-paranormal operators acting on the complex Hilbert space $\mathscr{H}$. The class of $(m,n)^*$-paranormal operators is characterized in terms of the Radon–Nikodym derivative of the measure λF^{-1} with respect to λ. Moreover, we discuss the conditions under which the classes of composition operators, weighted composition operators, multiplicative composition operators are $(m,n)^*$-paranormal.

12.1 INTRODUCTION

Let $\mathscr{H}$ be an infinite-dimensional complex Hilbert space and $B(\mathscr{H})$ denotes the C^*-algebra of all bounded linear operators on $\mathscr{H}$. Recall that an operator F is $(m,n)^*$-paranormal if $||F^*x||^{n+1} \leq m||F^{n+1}x|| \, ||x||^n$ for all $x \in \mathscr{H}$, where n is a positive integer and m is a positive real number [1].

Let (X,Σ,λ) be a sigma-finite measure space and T be a measurable transformation from X into itself. In this paper, $L^2(X,\Sigma,\lambda)$ is denoted by $L^2(\lambda)$. The paper deals with characterizations of $(m,n)^*$-paranormal operators on a complex Hilbert space $\mathscr{H}$.

The paper characterizes the class of $(m,n)^*$-paranormal operators in terms of the Radon-Nikodym derivative of the measure λF^{-1} with respect to λ. This means that the operators' behavior can be related to this derivative, possibly leading to a better understanding of their properties and spectral behavior.

DOI: 10.1201/9781032634142-12

The paper discusses conditions under which composition operators are $(m,n)^*$-paranormal. Composition operators are a specific type of operators defined as $C_F f = f \circ F$, $f \in L^2(\lambda)$, on $L^2(\lambda)$. The operator T induces a composition operator C_F on $L^2(\lambda)$ if

- The measure $\lambda \circ F^{-1}$ is absolutely continuous with respect to λ, and
- The Radon-Nikodym derivative $\frac{d(\lambda F^{-1})}{d\lambda}$ is essentially bounded.

The adjoint composition operator is given by $C_F^* f = h.E(f) \circ F^{-1}$.

- $C_F^k f = f \circ F^k$
- $C_F^{*k} f = h_k.E(f) \circ F^{-k}$
- $C_F^* C_F f = hf$
- $C_F C_F^* f = h \circ FE(f)$ for each $f \in L^2$.

For the complex valued Σ- measurable u on X, the weighted composition operator $W = W_{(\lambda,F)}$ is a linear transformation acting on $L^2(\lambda)$ is defined by $Wf = u.f \circ F$ and its adjoint is $W^* f = h.E(u.f) \circ F^{-1}$.

- $W^k f = u_k.f \circ F^k$
- $W^{*k} f = h_k.E(u_k.f) \circ F^{-k}$
- $W^* W f = h.E(u^2) \circ F^{-1}$
- $WW^* f = u(h \circ F).E(u.f)$ for each $f \in L^2$.

The multiplicative composite operator on $L^2(\lambda)$ is given by $M_{u,F}(f) = u \circ F.f \circ F$. Also, its adjoint is defined by $M_{u,F}^* f = u.h.E(f) \circ F^{-1}$.

- $M_{u,F}^k f = \prod_{i=1}^k (u \circ F^i).f \circ F^k$
- $M_{u,F}^* f = u.h.E(f) \circ F^{-1}$
- $M_{u,F}^{*k} f = u.h.\prod_{i=1}^k (E(u.h \circ F^{-i})).E(f) \circ F^{-k}$
- $M_{u,F}^* M_{u,F} f = u.h.E(u \circ F) \circ F^{-1}.f$
- $M_{u,F} M_{u,F}^* f = u \circ F.[u \circ F.h \circ F.E(f)]$ for each $f \in L^2$.

In the article [2], Senthilkumar et al. delved into weighted composition involving quasi-paranormal operators. Quasi-paranormal operators serve as a broader category encompassing paranormal operators, which, in turn, extends the concept of normal operators. The authors have explored the behavior of the weighted composition of these operators and investigated any associated properties.

Meanwhile, [3] presents the work of Veluchamy et al., who concentrated on k-quasi-P-normal composition, weighted composition, and composite multiplicative within a complex Hilbert space. The authors have defined and scrutinized the class of k-quasi-P-normal operators, likely connected to k-quasi-paranormal operators, and thoroughly examined their composition with other operators, including weighted versions.

In [4], Cowen et al. This paper establishes a correlation between subnormality of composition operators on the Hardy space $\mathscr{H}^2$ and the Denjoy-Wolff point. Subnormality is a well-known property of operators, and the Denjoy-Wolff point is related to the behavior of certain classes of holomorphic functions. The authors likely investigate how these two concepts are connected for composition operators on $\mathscr{H}^2$.

Many other authors [6, 7, 4, 8, 9, 5] cover a variety of topics related to composition operators, weighted composition operators, or multiplicative composition operators on different spaces. The authors could investigate a wide range of properties, including spectral theory, ergodic properties, and connections to other areas of mathematics.

12.2 $(M,N)^*$-PARANORMAL COMPOSITION OPERATORS

Proposition 12.1 *[1, Theorem 2.1] An operator F is $(m,n)^*$-paranormal if and only if*

$$m^{\frac{2}{n+1}} F^{*n+1} F^{n+1} - (n+1)x^n F F^* + m^{\frac{2}{n+1}} n a^{n+1} I \geq 0,$$

for each $x > 0$.

Theorem 12.1 *Let $C_F \in B(L^2)$ and x is any positive number, then the following conditions are identical*

1. C_F *is $(m,n)^*$-paranormal.*
2. $m^{\frac{1}{n+1}} \| \sqrt{h_n E(h) \circ F^{-n} f} \| \geq (n+1)^{\frac{1}{2}} a^{\frac{n}{2}} \| \sqrt{h \circ FE(f)} \| - m^{\frac{1}{n+1}} n^{\frac{1}{2}} x^{\frac{n+1}{2}} \|f\|.$
3. $m^{\frac{2}{n+1}} h_n E(h) \circ F^{-n} \geq (n+1)x^n (h \circ FE(f)) - m^{\frac{2}{n+1}} n x^{n+1}.$
4. $m^{\frac{2}{n+1}} h_{n-1} h \circ F^{-(n-1)} E(h) \circ F^{-n} \geq (n+1)a^n (h \circ FE(f)) - m^{\frac{2}{n+1}} n x^{n+1}.$
5. $m^{\frac{2}{n+1}} h.h_{m-1} \circ F^{-1} E(h) \circ F^{-n} \geq (n+1)a^n (h \circ FE(f)) - m^{\frac{2}{n+1}} n x^{n+1},$

Proof (1) $\Longrightarrow$ (2): *Since C_F is $(m,n)^*$-paranormal, by Proposition 12.1, for each $x > 0$, we have*

$$m^{\frac{2}{n+1}} C_F^{*n+1} C_F^{n+1} - (n+1)x^n C_F C_F^* + m^{\frac{2}{n+1}} n a^{n+1} \geq 0,$$

that is,

$$m^{\frac{2}{n+1}} C_F^{*n+1} C_F^{n+1} \geq (n+1)a^n C_F C_F^* - m^{\frac{2}{n+1}} n x^{n+1}.$$

So, we have

$$m^{\frac{2}{n+1}} \left\langle (C_F^{*n+1} C_F^{n+1}) f, f \right\rangle \geq (n+1)x^n \left\langle (C_F C_F^*) f, f \right\rangle - m^{\frac{2}{n+1}} n x^{n+1} \left\langle f, f \right\rangle. \tag{12.1}$$

Consider

$$\begin{aligned} (C_F^{*n+1} C_F^{n+1}) f &= C_F^{*n} (C_F^* C_F) C_F^n f \\ &= C_F^{*n} (C_F^* C_F)(f \circ F^n) \\ &= C_F^{*n} (h f \circ F^n) \\ &= h_n E(h f \circ F^n) \circ F^{-n} \\ &= h_n E(h) \circ F^{-n} . E(f) \circ F^n \circ F^{-n} \\ &= h_n E(h) \circ F^{-n} . f \end{aligned} \tag{12.2}$$

and

$$C_F C_F^* f = h \circ FE(f). \tag{12.3}$$

From (12.1), (12.2), *and* (12.3), *we obtain*

$$m^{\frac{2}{n+1}} \left\langle h_n E(h) \circ F^{-n} f, f \right\rangle \geq (n+1)x^n \left\langle h \circ FE(f), f \right\rangle - m^{\frac{2}{n+1}} n x^{n+1} \left\langle f, f \right\rangle,$$

that is,

$$m^{\frac{2}{n+1}} \| \sqrt{h_n E(h) \circ F^{-n} f} \|^2 \geq (n+1)x^n \| \sqrt{h \circ FE(f)} \|^2 - m^{\frac{2}{n+1}} n x^{n+1} \|f\|^2, \tag{12.4}$$

that is,

$$m^{\frac{1}{n+1}} \| \sqrt{h_n E(h) \circ F^{-n} f} \| \geq (n+1)^{\frac{1}{2}} x^{\frac{n}{2}} \| \sqrt{h \circ FE(f)} \| - m^{\frac{1}{n+1}} n^{\frac{1}{2}} x^{\frac{n+1}{2}} \|f\|.$$

(2) $\Longrightarrow$ (3): *From* (12.4), *we have*

$$m^{\frac{2}{n+1}} \left\langle h_n E(h) \circ F^{-n} f, f \right\rangle \geq (n+1)x^n \left\langle h \circ FE(f), f \right\rangle - m^{\frac{2}{n+1}} n x^{n+1} \left\langle f, f \right\rangle,$$

that is,

$$m^{\frac{2}{n+1}} h_n E(h) \circ F^{-n} \geq (n+1)x^n (h \circ FE) - m^{\frac{2}{n+1}} n x^{n+1}.$$

(3) $\implies$ (4): *Consider*

$$m^{\frac{2}{n+1}} h_n E(h) \circ F^{-n} \geq (n+1)x^n h - m^{\frac{2}{n+1}} n x^{n+1} \tag{12.5}$$

and

$$h_n = \lambda F^{-n}(B). \tag{12.6}$$

We have

$$\begin{aligned} \lambda F^{-n}(B) =& \lambda^{-1}(F^{-(n-1)}(B)) \\ =& \int_{F^{-(n-1)}} n d\lambda \\ =& h_{n-1} h \circ F^{-(n-1)}. \end{aligned} \tag{12.7}$$

From (12.6) *and* (12.7)*, we get*

$$h_n = h_{n-1} h \circ F^{-(n-1)}. \tag{12.8}$$

Now, substitute (12.8) *in* (12.5)*,*

$$m^{\frac{2}{n+1}} h_{n-1} h \circ F^{-(n-1)} E(h) \circ F^{-n} \geq (n+1)x^n (h \circ FE) - m^{\frac{2}{n+1}} n x^{n+1}.$$

(3) $\implies$ (5): *Observe that*

$$m^{\frac{2}{n+1}} h_n E(h) \circ F^{-n} \geq (n+1)x^n (h \circ FE) - m^{\frac{2}{n+1}} n x^{n+1} \tag{12.9}$$

and

$$h_n = \lambda F^{-n}(B). \tag{12.10}$$

We have

$$\begin{aligned} \lambda F^{-n}(B) =& \lambda(F^{-(n-1)}(F^{-1}B)) \\ =& \int_{F^{-B}} h_{n-1} d\lambda \\ =& h.h_{m-1} \circ F^{-1}. \end{aligned} \tag{12.11}$$

From (12.10) *and* (12.11)*, we obtain*

$$h_n = h.h_{m-1} \circ F^{-1}. \tag{12.12}$$

Now, substitute (12.12) *in* (12.9)*,*

$$m^{\frac{2}{n+1}} h.h_{m-1} \circ F^{-1} E(h) \circ F^{-n} \geq (n+1)x^n (h \circ FE) - m^{\frac{2}{n+1}} n x^{n+1}.$$

(5) $\implies$ (1) :
Consider that

$$m^{\frac{2}{n+1}} h.h_{m-1} \circ F^{-1} E(h) \circ F^{-n} \geq (n+1)x^n (h \circ FE) - m^{\frac{2}{n+1}} n x^{n+1}, \tag{12.13}$$

'holds. Since

$$h_n = h.h_{m-1} \circ F^{-1}.$$

Therefore, (12.13) *becomes*

$$\begin{aligned} & m^{\frac{2}{n+1}} h_n E(h) \circ F^{-n} \geq (n+1)x^n (h \circ FE) - m^{\frac{2}{n+1}} n x^{n+1}, \\ & m^{\frac{2}{n+1}} \langle h_n E(h) \circ F^{-n} f, f \rangle \geq (n+1)x^n \langle h \circ FE(f), f \rangle - m^{\frac{2}{n+1}} n x^{n+1} \langle f, f \rangle \end{aligned} \tag{12.14}$$

By definition of composition operator, we have

$$\begin{aligned} & (C_F^{*n+1} C_F^{n+1}) f = h_n E(h) \circ F^{-n}.f \\ & C_F C_F^* f = h \circ FE(f). \end{aligned} \tag{12.15}$$

Therefore, from (12.14) *and* (12.15), *we get*

$$m^{\frac{2}{n+1}}\left\langle (C_F^{*n+1}C_F^{n+1})f,f\right\rangle \geq (n+1)x^n\left\langle (C_FC_F^*)f,f\right\rangle - m^{\frac{2}{n+1}}nx^{n+1}\left\langle f,f\right\rangle. \tag{12.16}$$

Hence C_F is $(m,n)^*$ paranormal operator.
This proves the identical relations.

Theorem 12.2 *For each positive number x, an operator C_F^* is $(m,n)^*$-paranormal if and only if*

$$m^{\frac{2}{n+1}}\left\langle h_{n+1}\circ F^{n+1}.E(f),f\right\rangle \geq (n+1)x^n\left\langle hF,f\right\rangle - m^{\frac{2}{n+1}}nx^{n+1}\left\langle f,f\right\rangle, \tag{12.17}$$

Proof *Let C_F^* be $(m,n)^*$-paranormal. Then by Proposition 12.1, for each $x>0$, we have*

$$m^{\frac{2}{n+1}}C_F^{n+1}C_F^{*n+1}-(n+1)x^nC_F^*C_F+m^{\frac{2}{n+1}}nx^{n+1}\geq 0,$$

that is,

$$m^{\frac{2}{n+1}}C_F^{n+1}C_F^{*n+1}\geq (n+1)x^nC_F^*C_F-m^{\frac{2}{n+1}}nx^{n+1},$$

that is,

$$m^{\frac{2}{n+1}}\left\langle (C_F^{n+1}C_F^{*n+1})f,f\right\rangle \geq (n+1)x^n\left\langle (C_F^*C_F)f,f\right\rangle - m^{\frac{2}{n+1}}nx^{n+1}\left\langle f,f\right\rangle. \tag{12.18}$$

Observe that

$$(C_F^{n+1}C_F^{*n+1})f = h_{n+1}\circ F^{n+1}.E(f), \quad C_F^*C_Ff = hf. \tag{12.19}$$

Substitute (12.19) in (12.18),

$$m^{\frac{2}{n+1}}\left\langle h_{n+1}\circ F^{n+1}.E(f),f\right\rangle \geq (n+1)x^n\left\langle hf,f\right\rangle - m^{\frac{2}{n+1}}nx^{n+1}\left\langle f,f\right\rangle.$$

Conversely, let (12.17) *holds.*
Consider

$$h_{n+1}\circ F^{n+1}.E(f) = (C_F^{n+1}C_F^{*n+1})f, \quad hf = C_F^*C_Ff. \tag{12.20}$$

Substitute (12.20) *in* (12.17), *we obtain*

$$m^{\frac{2}{n+1}}\left\langle (C_F^{n+1}C_F^{*n+1})f,f\right\rangle \geq (n+1)x^n\left\langle C_F^*C_Ff,f\right\rangle - m^{\frac{2}{n+1}}nx^{n+1}\left\langle f,f\right\rangle,$$

that is,

$$m^{\frac{2}{n+1}}(C_F^{n+1}C_F^{*n+1})\geq (n+1)x^nC_F^*C_F-m^{\frac{2}{n+1}}nx^{n+1}.$$

Hence, C_F^* is $(m,n)^*$-paranormal.

12.3 MULTIPLICATIVE COMPOSITION OPERATORS

Theorem 12.3 *For each positive number x, the multiplicative composition operator $M_{u,F}$ is $(m,n)^*$-paranormal operator if and only if*

$$m^{\frac{2}{n+1}}\|\prod_{i=1}^{n+1}(u\circ F^i).f\circ F^{n+1}\|^2 \geq (n+1)x^n\|u.h.E(f)\circ F^{-1}\|^2 - m^{\frac{2}{n+1}}nx^{n+1}\|f\|^2. \tag{12.21}$$

Proof *Let $M_{u,F}$ be $(m,n)^*$-paranormal. Then for each $a>0$,*

$$m^{\frac{2}{n+1}}M_{u,F}^{*n+1}M_{u,F}^{n+1}-(n+1)x^nM_{u,F}^*M_{u,F}+m^{\frac{2}{n+1}}nx^{n+1}\geq 0,$$

that is,

$$m^{\frac{2}{n+1}}M_{u,FT}^{*n+1}M_{u,F}^{n+1}\geq (n+1)x^nM_{u,F}M_{u,F}^*-m^{\frac{2}{n+1}}nx^{n+1},$$

that is,

$$m^{\frac{2}{n+1}}\left\langle (M_{u,F}^{*n+1}M_{u,F}^{n+1})f,f\right\rangle \geq (n+1)x^n\left\langle (M_{u,F}M_{u,F}^*)f,f\right\rangle - m^{\frac{2}{n+1}}nx^{n+1}\langle f,f\rangle, \tag{12.22}$$

that is,

$$m^{\frac{2}{n+1}}\|M_{u,F}^{n+1}f\|^2\geq (n+1)x^n\|M_{u,F}^*f\|^2-m^{\frac{2}{n+1}}nx^{n+1}\|f\|^2.$$

Since

$$M_{u,F}^{n+1}f=\prod_{i=1}^{n+1}(u\circ F^i).f\circ F^{n+1}, \quad M_{u,F}^*f=u.h.E(f)\circ F^{-1}. \tag{12.23}$$

Substitute (12.23) *in* (12.22), *so we get*

$$m^{\frac{2}{n+1}}\|\prod_{i=1}^{n+1}(u\circ F^i).f\circ F^{n+1}\|^2 \geq (n+1)x^n\|u.h.E(f)\circ F^{-1}\|^2 - m^{\frac{2}{n+1}}nx^{n+1}\|f\|^2.$$

Conversely, let (12.21) *holds.*

Since

$$M_{u,F}^{n+1}f=\prod_{i=1}^{n+1}(u\circ F^i).f\circ F^{n+1}, \quad M_{u,F}^*f=u.h.E(f)\circ F^{-1}. \tag{12.24}$$

Substitute (12.24) *in* (12.21),

$$m^{\frac{2}{n+1}}\|M_{u,F}^{n+1}f\|^2\geq (n+1)x^n\|M_{u,F}^*f\|^2-m^{\frac{2}{n+1}}nx^{n+1}\|f\|^2,$$

that is,

$$m^{\frac{2}{n+1}}\left\langle (M_{u,F}^{*n+1}M_{u,F}^{n+1})f,f\right\rangle \geq (n+1)x^n\left\langle (M_{u,F}M_{u,F}^*)f,f\right\rangle - m^{\frac{2}{n+1}}nx^{n+1}\langle f,f\rangle,$$

that is,

$$m^{\frac{2}{n+1}}M_{u,F}^{*n+1}M_{u,F}^{n+1}-(n+1)x^nM_{u,F}M_{u,F}^*+m^{\frac{2}{n+1}}nx^{n+1}\geq 0,$$

for each $x>0$. Thus, $M_{u,F}$ is $(m,n)^*$-paranormal.

Theorem 12.4 *For each positive number x, the adjoint multiplicative composition operator $M^*_{u,F}$ is $(m,n)^*$-paranormal operator if and only if*

$$m^{\frac{2}{n+1}} \|u.h. \prod_{i=1}^{n+1} (E(u.h \circ F^{-i})).E(f) \circ F^{-n+1}\|^2 \geq (n+1)x^n \|u \circ F.f \circ F\|^2 - m^{\frac{2}{n+1}} nx^{n+1} \|f\|^2. \tag{12.25}$$

Proof *Assume that $M^*_{u,F}$ is $(m,n)^*$ paranormal. Then for each $a > 0$,*

$$m^{\frac{2}{n+1}} M^{n+1}_{u,F} M^{*n+1}_{u,F} - (n+1)x^n M^*_{u,F} M_{u,F} + m^{\frac{2}{n+1}} nx^{n+1} \geq 0,$$

that is,

$$m^{\frac{2}{n+1}} M^{n+1}_{u,F} M^{*n+1}_{u,F} \geq (n+1)x^n M^*_{u,F} M_{u,F} - m^{\frac{2}{n+1}} nx^{n+1},$$

that is,

$$m^{\frac{2}{n+1}} \left\langle (M^{n+1}_{u,F} M^{*n+1}_{u,F}) f, f \right\rangle \geq (n+1)x^n \langle (M^*_{u,F} M_{u,F}) f, f \rangle - m^{\frac{2}{n+1}} nx^{n+1} \langle f, f \rangle, \tag{12.26}$$

that is,

$$m^{\frac{2}{n+1}} \|M^{*n+1}_{u,F} f\|^2 \geq (n+1)x^n \|M_{u,F} f\|^2 - m^{\frac{2}{n+1}} nx^{n+1} \|f\|^2.$$

Consider

$$M^{*n+1}_{u,F} f = u.h. \prod_{i=1}^{n+1} (E(u.h \circ F^{-i})).E(f) \circ F^{-n+1}, \quad M_{u,F} f = u \circ F.f \circ F. \tag{12.27}$$

Substitute (12.27) *in* (12.26), *so we obtain*

$$m^{\frac{2}{n+1}} \|u.h. \prod_{i=1}^{n+1} (E(u.h \circ F^{-i})).E(f) \circ F^{-n+1}\|^2 \geq (n+1)x^n \|u \circ F.f \circ F\|^2 - m^{\frac{2}{n+1}} nx^{n+1} \|f\|^2. \tag{12.28}$$

Conversely, let (12.25) *holds.*

Consider

$$M^{*n+1}_{u,F} f = u.h. \prod_{i=1}^{n+1} (E(u.h \circ F^{-i})).E(f) \circ F^{-n+1}, \quad M_{u,F} f = u \circ F.f \circ F. \tag{12.29}$$

Substitute (12.29) *in* (12.25),

$$m^{\frac{2}{n+1}} \|M^{*n+1}_{u,F} f\|^2 \geq (n+1)x^n \|M_{u,F} f\|^2 - m^{\frac{2}{n+1}} nx^{n+1} \|f\|^2,$$

that is,

$$m^{\frac{2}{n+1}} \left\langle (M^{n+1}_{u,F} M^{*n+1}_{u,F}) f, f \right\rangle \geq (n+1)x^n \langle (M^*_{u,F} M_{u,F}) f, f \rangle - m^{\frac{2}{n+1}} nx^{n+1} \langle f, f \rangle,$$

that is,

$$m^{\frac{2}{n+1}} M^{n+1}_{u,F} M^{*n+1}_{u,F} - (n+1)x^n M^*_{u,F} M_{u,F} + m^{\frac{2}{n+1}} nx^{n+1} \geq 0,$$

for each $a > 0$. Thus, $M^*_{u,F}$ is $(m,n)^*$-paranormal.

12.4 WEIGHTED COMPOSITION OPERATORS

Theorem 12.5 *For each positive number x, the weighted composition operator W is $(m,n)^*$-paranormal if and only if*

$$m^{\frac{2}{n+1}}\|u_{n+1}.f\circ F^{n+1}\|^2 \geq (n+1)x^n\|h.E(u.f)\circ F^{-1}\|^2 - m^{\frac{2}{n+1}}nx^{n+1}\|f\|^2, \tag{12.30}$$

for each $f\in L^2$ and $a>0$.

Proof *Assume that W is $(m,n)^*$-paranormal. Then, for each $a>0$,*

$$m^{\frac{2}{n+1}}W^{*n+1}W^{n+1} - (n+1)x^nWW^* + m^{\frac{2}{n+1}}nx^{n+1} \geq 0,$$

that is,

$$m^{\frac{2}{n+1}}W^{*n+1}W^{n+1} \geq (n+1)x^nWW^* - m^{\frac{2}{n+1}}nx^{n+1},$$

that is,

$$m^{\frac{2}{n+1}}\left\langle (W^{*n+1}W^{n+1})f,f\right\rangle \geq (n+1)x^n\left\langle (WW^*)f,f\right\rangle - m^{\frac{2}{n+1}}nx^{n+1}\left\langle f,f\right\rangle,$$

that is,

$$m^{\frac{2}{n+1}}\|W^{n+1}f\|^2 \geq (n+1)x^n\|W^*f\|^2 - m^{\frac{2}{n+1}}nx^{n+1}\|f\|^2. \tag{12.31}$$

Consider

$$\begin{aligned} W^{n+1}f &= u_{n+1}.f\circ F^{n+1}, \\ W^*f &= h.E(u.f)\circ F^{-1}. \end{aligned} \tag{12.32}$$

Substitute (12.32) *in* (12.31), *we get*

$$m^{\frac{2}{n+1}}\|u_{n+1}.f\circ F^{n+1}\|^2 \geq (n+1)x^n\|h.E(u.f)\circ F^{-1}\|^2 - m^{\frac{2}{n+1}}nx^{n+1}\|f\|^2. \tag{12.33}$$

Conversely, let (12.30) holds.
Consider

$$\begin{aligned} W^{n+1}f &= u_{n+1}.f\circ F^{n+1}, \\ W^*f &= h.E(u.f)\circ F^{-1}. \end{aligned} \tag{12.34}$$

Substitute (12.34) *in* (12.30),

$$m^{\frac{2}{n+1}}\|W^{n+1}f\|^2 \geq (n+1)x^n\|W^*f\|^2 - m^{\frac{2}{n+1}}nx^{n+1}\|f\|^2,$$

that is,

$$m^{\frac{2}{n+1}}\left\langle (W^{*n+1}W^{n+1})f,f\right\rangle \geq (n+1)x^n\left\langle (WW^*)f,f\right\rangle - m^{\frac{2}{n+1}}nx^{n+1}\left\langle f,f\right\rangle,$$

that is,

$$m^{\frac{2}{n+1}}W^{*n+1}W^{n+1} \geq (n+1)x^nWW^* - m^{\frac{2}{n+1}}nx^{n+1}.$$

This completes the converse part.

Theorem 12.6 *For each positive number x, the adjoint weighted composition operator* W^* *is* $(m,n)^*$*-paranormal if and only if*

$$m^{\frac{2}{n+1}}\|h_{n+1}E(u_{n+1}.f)\circ F^{-(n+1)}\|^2 \geq (n+1)x^n\|u.f\circ F\|^2 - m^{\frac{2}{n+1}}nx^{n+1}\|f\|^2, \tag{12.35}$$

for each $f \in L^2$.

Proof *Let* W^* *be* $(m,n)^*$*-paranormal. Then for each* $a > 0$,

$$m^{\frac{2}{n+1}}W^{n+1}W^{*n+1} - (n+1)x^nWW^* + m^{\frac{2}{n+1}}nx^{n+1} \geq 0,$$

that is,

$$m^{\frac{2}{n+1}}W^{n+1}W^{*n+1} \geq (n+1)x^nW^*W - m^{\frac{2}{n+1}}nx^{n+1},$$

that is,

$$m^{\frac{2}{n+1}}\left\langle (W^{n+1}W^{*n+1})f,f\right\rangle \geq (n+1)x^n\left\langle (W^*W)f,f\right\rangle - m^{\frac{2}{n+1}}nx^{n+1}\left\langle f,f\right\rangle,$$

that is,

$$m^{\frac{2}{n+1}}\|W^{*n+1}f\|^2 \geq (n+1)x^n\|Wf\|^2 - m^{\frac{2}{n+1}}nx^{n+1}\|f\|^2. \tag{12.36}$$

Consider

$$\begin{aligned} W^{*n+1}f &= h_{n+1}E(u_{n+1}.f)\circ F^{-(n+1)}, \\ Wf &= u.f\circ F. \end{aligned} \tag{12.37}$$

Substitute (12.37) *in* (12.36),

$$m^{\frac{2}{n+1}}\|h_{n+1}E(u_{n+1}.f)\circ F^{-(n+1)}\|^2 \geq (n+1)x^n\|u.f\circ F\|^2 - m^{\frac{2}{n+1}}nx^{n+1}\|f\|^2.$$

Conversely, let (12.35) holds.
Since

$$\begin{aligned} W^{*n+1}f &= h_{n+1}E(u_{n+1}.f)\circ F^{-(n+1)}, \\ Wf &= u.f\circ F. \end{aligned} \tag{12.38}$$

Substitute (12.38) *in* (12.35),

$$m^{\frac{2}{n+1}}\|W^{*n+1}f\|^2 \geq (n+1)x^n\|Wf\|^2 - m^{\frac{2}{n+1}}nx^{n+1}\|f\|^2,$$

that is,

$$m^{\frac{2}{n+1}}\left\langle (W^{n+1}W^{*n+1})f,f\right\rangle \geq (n+1)x^n\left\langle (W^*W)f,f\right\rangle - m^{\frac{2}{n+1}}nx^{n+1}\left\langle f,f\right\rangle,$$

that is,

$$m^{\frac{2}{n+1}}W^{n+1}W^{*n+1} \geq (n+1)x^nW^*W - m^{\frac{2}{n+1}}nx^{n+1}.$$

Hence, W^* *is* $(m,n)^*$*-paranormal.*

CONCLUSION

In conclusion, this paper establishes significant characterizations for the class of $(m,n)^*$-paranormal operators in the context of complex Hilbert space $\mathscr{H}$. The central result involves a characterization of $(m,n)^*$-paranormal operators by relating them to the Radon–Nikodym derivative of the measure λF^{-1} with respect to λ. This characterization provides insight into the nature of $(m,n)^*$-paranormal operators and their connection to measure theory. Furthermore, the paper explores the conditions under which various subclasses of operators, including composition operators, weighted composition operators, and multiplicative composition operators, belong to the class of $(m,n)^*$-paranormal operators. This analysis expands our understanding of the $(m,n)^*$-paranormal property and its applicability to specific operator classes.

REFERENCES

1. Dharmarha, P. and Ram, S.: (m,n)-paranormal operators and $(m,n)^*$-paranormal operators, Commun. Korean Math. Soc. Soc. **35**(1), 151–159. (2020) DOI : 10.4134/CKMS.c180452
2. Senthilkumar, D., Naik, P.M., and Santhi, R.: Weighted composition of quasi-paranormal operator, Far East Journal of Mathematical Sciences **72** (2013), no. 2, 369–383.
3. Veluchamy, T.and Manikandan, K.M.: k-quasi-P-normal composition, weighted composition and composite multiplication on the complex hilbert space, Int. J. Pure. Appli. Maths. **119** (2018), no. 2, 14239–14266.
4. Cowen, C. and Kriete, T.L.: Subnormality and Composition Operators on H^2, J. Funct. Anal. **81**(1988), 298-319.
5. Kour, B. and Ram, S.: (m, n)-Paranormal Composition Operators. In Mathematical Analysis and Applications. Springer Proceedings in Mathematics and Statistics, vol 381. Springer, Singapore (2021) https://doi.org/10.1007/978-981-16-8177-6˙13
6. Senthilkumar, D. and Thirugnanasambandam, K.: M-Paranormal and * Paranormal Composition Operators on Weighted Hardy Space, Int. J. Contemp. Math. Sciences **5** (2010), no. 56, 2793–2799.
7. Senthil, S., Thangaraju, P. and Kumar, D.C.: k-*Paranormal, k-Quasi-*paranormal and (n,k)-Quasi-*paranormal Composite Multiplication Operator on L^2-spaces, British Journal of Mathematics and Computer Science **11** (2015), no. 6, 1–15.
8. Senthilkumar, D. and Selvi, P.T.: Weighted Composition of m-Quasi k-Paranormal Operators, Int. J. Math. And Appl. **7**, no. 1, 1–8 (2019) http://ijmaa.in/index.php/ijmaa/article/view/262
9. Sivamani, N.: Posinormal and * Paranormal Composition Operators on the Fock Space, International Journal of Innovative Science, Engineering and Technology **2** (2012), no. 12.

13 On the Local Convergence of a Fifth-Order Method for Solving a System of Non-Linear Equations

Jinny Ann John
Department of Mathematics, Puducherry Technological University, Pondicherry, India

Jayakumar Jayaraman
Department of Mathematics, Puducherry Technological University, Pondicherry, India

CONTENTS

Methods of high convergence order are important in the field of Computational Mathematics, since they generate sequences which converges to the zero of a non-linear equation. In order to prove the order of convergence of these methods, Taylor series expansion is used, which may require the existence of derivatives of higher order that are not present in the method. Hence, the existing results may not assure the convergence of the method in cases where the high-order derivatives do not exist. But the method may converge. In this article, we consider a fifth-order iterative scheme for solving a system of non-linear equation in a Banach space domain. Using ω-continuity conditions and hypotheses based only on first derivative of the operator, we provide the local convergence analysis of the iterative scheme under consideration. This is in contrast to the existing works where derivatives up to order six are required to prove the convergence of the method. As a result, we are able to enlarge the utility of the method under consideration. This methodology does not depend on the method and hence can be extended to other methods in a similar manner. Furthermore, we also provide results for finding computable radii of the convergence domain, estimates on error bounds and also results on proving the uniqueness of the solution. Finally, some numerical illustrations are provided where the convergence conditions are verified.

DOI: 10.1201/9781032634142-13

13.1 INTRODUCTION

As technology advances day by day, plenty of highly complex scientific and engineering problems emerges. With the help of tools in Mathematical modelling, a streamlined approach can be made which will significantly reduce these problems into a non-linear equation of the form

$$T(x) = 0, \tag{13.1}$$

where $T : O \subset X_1 \to X_2$ is a continuously differentiable non-linear operator betwixt two Banach spaces X_1 and X_2 and O is an open, non-empty convex subset of X_1.

Scientists and researches have strived for long to find plausible solutions to these non-linear equations. A direct solution is not possible in most of the cases. This is where the role of iterative methods comes into action. These iterative methods give rise to a iterative sequence which in turn converges to the zero of the non-linear equation. Newton-Raphson method is the most celebrated among the iterative processes which converges to the root quadratically. On improvising the order of convergence, many higher-order multi-point methods were then formulated depending upon the need [2, 4, 9, 10]. The order of convergence of these methods were proved using the Taylor series expansion which involves the computation of higher-order derivatives. In most of the cases, these derivatives of higher order are not present in the method and also may not exist which turned out to be an inevitable disadvantage. Additionally, a wild guess is made to choose the initial point and no results on the domain of convergence were discussed in these type of studies. Furthermore, error estimates and uniqueness of the solution were also not specified.

Contemporary researches then started focussing on resolving the shortcomings of the traditional Taylor series method. As a result of which commenced the study of local and semi-local convergence behavior of iterative methods which are summarised in [1, 3, 5, 6, 8]. These type of studies examines the convergence domain, error estimates and also uniqueness of the solution. Resultant of these studies are relevant as they enlarge working range of the iterative methods.

This paper deals with the development of results concerning the local behavior of a fifth-order iterative scheme primarily found in [9]. We consider the method in a Banach space setting as follows

$$\begin{aligned}
y_j &= x_j - \frac{T(x_j)}{T'(x_j)} \\
x_{j+1} &= y_j - \frac{T(y_j)}{T'(y_j)} G_j, \\
&\text{where } G_j = g_1 + g_2\eta(x_j, y_j) + g_3\eta(x_j, y_j)^2, \\
\eta(x_j, y_j) &= \frac{T'(y_j)}{T'(x_j)}, \\
g_1 &= \frac{9}{8}, g_2 = \frac{1}{8} \text{ and } g_3 = -\frac{1}{4} \text{ for } x_0 \in O \text{ and } \forall j = 0, 1, 2, \cdots.
\end{aligned} \tag{13.2}$$

The convergence order of method (13.2) was established assuming that the sixth derivative (at least) of the operator T exists. As a result, applicability of these iterative methods is limited. We can observe this by defining T on $O = [-0.5, 1.5]$ by

$$T(p) = \begin{cases} p^3 \ln(p^2) + p^5 - p^4, & \text{if } p \neq 0 \\ 0, & \text{if } p = 0. \end{cases} \tag{13.3}$$

Now it is easy to notice that due to the unboundedness of T''', the results on convergence of (13.4) does not hold for this example. Also, these articles do not produce any formula for approximating the error $\|x_r - \upsilon\|$, the convergence region or the uniqueness and accurate location of the root υ. This encourages us to develop the ball convergence theorems and hence compare the convergence domains by considering assumptions only on T'. Our research provides important formulas for the estimation of $\|x_r - \upsilon\|$ and convergence radii. This study also discusses about an exact location and the uniqueness of υ.

The remaining contents of this article can be summarized in the following manner: Section 13.2 deals with local convergence analysis of the method (13.2). Numerical illustrations are placed in Section 13.3. Conclusions are stated in Section 13.4.

13.2 LOCAL CONVERGENCE ANALYSIS

In this section, we present the local convergence analysis of the method (13.2). Let $U(c,\mu)$ and $\bar{U}(c,\mu)$ denote open and closed balls in O, respectively, with centre c and radius $\mu > 0$. Next, we develop some real parameters and functions which in turn helps in proving the local convergence of the method under consideration.

Suppose ω, ω_0 and ω_1 are continuous and non-decreasing functions defined on the interval $[0,+\infty)$ such that $\omega(0) = \omega_0(0) = \omega_1(0) = 0$. Assume that S_0 is the smallest positive zero of $\omega_0(p) = 1$. We define the functions V_1 and M_1 on the interval $[0,S_0)$ by

$$V_1(p) = \frac{\int_0^1 \omega((1-\kappa)p)d\kappa}{1-\omega_0(p)}$$

and

$$M_1(p) = V_1(p) - 1.$$

Notice that $M_1(0) = -1 < 0$ and $\lim_{p \to S_0^-} M_1(p) = +\infty$. Hence, the intermediate value theorem guarantees the existence of the smallest zero α_1 of $M_1(p)$ in $(0,S_0)$. Therefore, $V_1(\alpha_1) = 1$ and $0 \leq V_1(p) < 1$ for $p \in [0,\alpha_1)$.

Assume that there exists continuous and non-decreasing functions V_2 and M_2 on the interval $[0,S_0)$ such that

$$V_2(p) = \omega_0(V_1(p)p)$$

and

$$M_2(p) = V_2(p) - 1.$$

Hence, again by intermediate value theorem we can conclude that $M_2(p)$ has a least root S_1 in $(0,S_0)$.

Subsequently, we define the functions V_3 and M_3 on $[0,S_1)$ by

$$V_3(p) = \frac{\int_0^1 \omega((1-\kappa)V_1(p)p)d\kappa}{1-V_2(p)} + \chi(p)\frac{\int_0^1 \omega_1(\kappa(V_1(p)p))d\kappa}{1-V_2(p)}$$

and

$$M_3(p) = V_3(p) - 1,$$

where $\chi(p) = |g_1 - 1| + |g_2|\frac{\omega_1(V_1(p)p)}{1-\omega_0(p)} + |g_3|(\frac{\omega_1(V_1(p)p)}{1-\omega_0(p)})^2$.

Now, we can observe that $M_3(0) = -1 < 0$ and $\lim_{p \to S_1^-} M_3(p) = +\infty$. So, by intermediate value theorem there exists a least zero α_2 of $M_3(p)$ in $(0,S_1)$.

Define

$$\alpha = min\{\alpha_1,\alpha_2\}. \tag{13.4}$$

Now, for each $p \in [0, \alpha)$ observe that the following holds:

$$0 \leq \omega_0(p) < 1, \tag{13.5}$$

$$0 \leq V_k(p) < 1, \ k = 1,2,3. \tag{13.6}$$

We shall prove that α is the radius of the convergence domain of the iterative method (13.2).
The subsequent conditions are employed in proving the local convergence.

(L_1) The function $T : O \to X_2$ is continuous and Frechet differentiable and let υ be a simple root of the equation (13.1).

(L_2) $\|T'(\upsilon)^{-1}(T'(q) - T'(\upsilon))\| \leq \omega_0(\|q - \upsilon\|)$ for all $q \in O$. Fix $O_0 = O \cap U(\upsilon, S_0)$.

(L_3) $\|T'(\upsilon)^{-1}(T'(q) - T'(t)\| \leq \omega(\|q - t\|)$ for each $q, t \in O_0$. Set $O_1 = O \cap \bar{U}(\upsilon, S)$.

(L_4) $\|T'(\upsilon)^{-1}T'(q)\| \leq \omega_1(\|q - \upsilon\|)$ for all $q \in O_1$.

(L_5) $\bar{U}(\upsilon, \alpha) \subset O$.

(L_6) If there exists $\alpha_3 > \alpha$ such that $\int_0^1 \omega_0(\kappa\alpha_3)d\kappa < 1$, then set $O_2 = O \cap \bar{U}(\upsilon, \alpha_3)$.

Theorem 13.1 *Assume that the conditions (L_1)-(L_5) hold and select a starting point $x_0 \in U(\upsilon, \alpha) - \{\upsilon\}$. Then, the subsequent conditions hold*

$$\{x_r\} \subset U(\upsilon, \alpha), \tag{13.7}$$

$$\|y_r - \upsilon\| \leq V_1(\|x_r - \upsilon\|)\|x_r - \upsilon\| \leq \|x_r - \upsilon\| < \alpha, \tag{13.8}$$

$$\|x_{r+1} - \upsilon\| \leq V_3(\|x_r - \upsilon\|)\|x_r - \upsilon\| \leq \|x_r - \upsilon\| \tag{13.9}$$

and

$$\lim_{n \to \infty} x_r = \upsilon, \tag{13.10}$$

where the radius α is as defined in (13.4) and the functions V_k are as formerly defined.

Proof *The relations (13.7)-(13.10) will be proved using the principle of mathematical induction. Let z be an arbitrary point such that $z \in U(\upsilon, \alpha) - \{\alpha\}$. Using (13.4), (13.5) and the conditions (L_1),(L_2), we have*

$$\|T'(\upsilon)^{-1}(T'(z) - T'(\upsilon))\| \leq \omega_0(\|z - \upsilon\|) \leq \omega_0(\alpha) < 1. \tag{13.11}$$

Now, the estimate (13.11) and the standard perturbation Lemma given by Banach for linear operators [7] asserts that $T'(z)^{-1} \in L(X_1, X_2)$ and

$$\|T'(z)^{-1}T'(\upsilon)\| \leq \frac{1}{1 - \omega_0(\|z - \upsilon\|)}. \tag{13.12}$$

Hence, observe that y_0 exists and using the first sub-step of the iterative scheme (13.2), we obtain

$$\begin{aligned} y_0 - \upsilon &= x_0 - \upsilon - T'(x_0)^{-1}T(x_0) \\ &= -T'(x_0)^{-1} \int_0^1 \left[T'(\upsilon + \kappa(x_0 - \upsilon)) - T'(x_0)\right] d\kappa(x_0 - \upsilon) \end{aligned} \tag{13.13}$$

Then, the identity (13.13), definition (13.4), the relation (13.6) (for k=1), condition(L_3) and equation (13.12)(for $z = x_0$) give

$$\begin{aligned} \|y_0 - \upsilon\| &\leq \frac{\int_0^1 \omega((1-\kappa)\|x_0 - \upsilon)d\kappa}{1 - \omega_0(\|x_0 - \upsilon\|)} \|x_0 - \upsilon\| \\ &\leq V_1(\|x_0 - \upsilon\|)\|x_0 - \upsilon\| \\ &\leq \|x_0 - \upsilon\| < \alpha. \end{aligned} \tag{13.14}$$

Consequently, the iterate $y_0 \in U(\upsilon,\alpha)$ *and also the relation (13.8) holds for* $r=0$.
Forthwith, making use of equation (13.12), we get that $T'(y_0)^{-1}$ *is invertible. Accordingly,* x_1 *exists and from the second sub-step of the method (13.2), we can write that*

$$\begin{aligned} x_1-\upsilon &= y_0-\upsilon-T'(y_0)^{-1}T(y_0)+(I-G_0)T'(y_0)^{-1}T(y_0) \\ &= -T'(y_0)^{-1}\int_0^1 \left[T'(\upsilon+\kappa(y_0-\upsilon))-T'(y_0)\right]d\kappa(y_0-\upsilon) \\ &\quad +(I-G_0)T'(y_0)^{-1}\int_0^1 T'(\upsilon+\kappa(y_0-\upsilon))d\kappa(y_0-\upsilon). \end{aligned} \tag{13.15}$$

Thereupon, by employing the identity (13.15), definition (13.4), the relation (13.6) (for k=3), condition (L_4) *and equation (13.12)(for* $z=x_0,y_0$*), we derive*

$$\begin{aligned} \|x_1-\upsilon\| &\le \left[\frac{\int_0^1 \omega((1-\kappa)\|y_0-\upsilon\|)d\kappa}{1-\omega_0(\|y_0-\upsilon\|)}+\chi(\|x_0-\upsilon\|)\frac{\int_0^1 \omega_1(\kappa\|y_0-\upsilon\|)}{1-\omega_0(\|y_0-\upsilon\|)}\right]\|y_0-\upsilon\| \\ &\le V_3(\|x_0-\upsilon\|)\|x_0-\upsilon\| \\ &\le \|x_0-\upsilon\|, \end{aligned} \tag{13.16}$$

which evinces the inequality (13.9) for $r=0$*, and also we get* $x_1\in U(\upsilon,\alpha)$*. In proving inequality (13.16), we also used the result*

$$\begin{aligned} \|I-G_0\| &\le |g_1-1|+|g_2|\frac{\omega_1(\|y_0-\upsilon\|)}{1-\omega_0(\|x_0-\upsilon\|)}+|g_3|\left(\frac{\omega_1(\|y_0-\upsilon\|)}{1-\omega_0(\|x_0-\upsilon\|)}\right)^2 \\ &\le \chi(\|x_0-\upsilon\|) \text{ (where } G_0 \text{ is as defined in (13.2)).} \end{aligned}$$

Next, by exchanging x_0,y_0,x_1 *by* x_r,y_r,x_{r+1} *in the above calculations, we complete the induction for the relations (13.7)-(13.9).*
Consequently, using the inequality

$$\|x_{r+1}-\upsilon\|\le Q\|x_r-\upsilon\|<\alpha,$$

where $Q=V_3(\|x_0-\upsilon\|)\in[0,1)$*, we can conclude that* $x_{r+1}\in U(\upsilon,\alpha)$ *and* $\lim_{r\to\infty}x_r=\upsilon$.

We now proceed to prove the uniqueness of the solution.

Proposition 13.1 *Presume that*

(a) *there exists a* $\delta\in U(\upsilon,\alpha_3)$ *such that* $T(\delta)=0$*;*

(b) *the condition* (L_2) *holds on* $U(\upsilon,\alpha_3)$*;*

(c) *the condition* (L_6) *holds.*

Then, the equation $T(x)=0$ *is uniquely soluble by* υ *in the region* O_2.

Proof *Let D be a linear operator and fix*

$$D=\int_0^1 T'(\upsilon+\kappa(\delta-\upsilon))d\kappa, \text{ for } \delta\in O_2 \text{ such that } T(\delta)=0.$$

Then, it follows by (a),(b) and (c) that

$$\begin{aligned} \|T'(\upsilon)^{-1}(D-T'(\upsilon)\| &\le \int_0^1 \omega_0(\kappa\|\delta-\upsilon\|)d\kappa \\ &\le \int_0^1 \omega_0(\kappa\alpha_3)d\kappa<1. \end{aligned}$$

As a result, D is invertible and $D^{-1} \in L(X_1, X_2)$. Consequently, using the invertibility of D and the identity

$$\upsilon - \delta = D^{-1}(T(\upsilon) - T(\delta)) = D^{-1}(0-0) = 0,$$

we can deduce that $\upsilon = \delta$.

13.2.1 REMARKS

We can easily calculate the convergence order without resorting to the traditional Taylor's series method, which involves the computation of higher-order derivatives using the following formulas:

$$\text{Computational Order of Convergence, } COC = \ln\left(\frac{\|x_{r+1} - \upsilon\|}{\|x_r - \upsilon\|}\right) / \ln\left(\frac{\|x_r - \upsilon\|}{\|x_{r-1} - \upsilon\|}\right)$$

and Approximate Computational Order of Convergence,

$$ACOC = \ln\left(\frac{\|x_{r+1} - x_r\|}{\|x_r - x_{r-1}\|}\right) / \ln\left(\frac{\|x_r - x_{r-1}\|}{\|x_{r-1} - x_{r-2}\|}\right).$$

13.3 EXAMPLES

Example 13.1 *Let $X_1 = X_2 = C[0,1]$ and $O = \bar{U}(0,1)$. Consider the non-linear Hammerstein type integral equation given by*

$$T(s)(p) = s(p) - 5\int_0^1 pts(t)^3 dt$$

where $s(p) \in C[0,1]$. Then, we have $\upsilon = 0$ is a solution. We find $\omega_0(p) = 7.5p$, $\omega(p) = 15p$ and $\omega_1(p) = 2$. The radii are obtained as $\alpha_1 = 0.0666667$ and $\alpha_2 = 0.152368$. Thus, using definition (13.4), we obtain the radius of the convergence domain as $\alpha = 0.0666667$.

Example 13.2 *Set $X_1 = X_2 = O = \mathbb{R}$. Define $T(x) = \sin x$. Then, we get $T'(x) = \cos x$. We have $\upsilon = 0$ is a solution. Accordingly, we acquire $\omega_0(p) = \omega(p) = p$ and $\omega_1(p) = 1$. Thence, we have the radii as $\alpha_1 = 0.666667$ and $\alpha_2 = 0.524271$. Therefore, by using definition (13.4), we get the radii of the domain of convergence as $\alpha = 0.524271$.*

Example 13.3 *Consider the system of differential equations with*

$$T_1'(p_1) = e^{p_1}, \quad T_2'(p_2) = (e-1)p_2 + 1, \quad T_3'(p_3) = 1$$

subject to the initial conditions $T_1(0) = T_2(0) = T_3(0) = 0$. Let $T = (T_1, T_2, T_3)$. Let $X_1 = X_2 = \mathbb{R}^3$ and $O = \bar{U}(0,1)$. Then $\upsilon = (0,0,0)^T$. Define the function T on O for $p = (p_1, p_2, p_3)^T$ as

$$T(p) = (e^{p_1} - 1, \frac{e-1}{2}p_2^2 + p_2, p_3)^T.$$

We get the Frechet derivative as

$$T'(p) = \begin{bmatrix} e^{p_1} & 0 & 0 \\ 0 & (e-1)p_2 + 1 & 0 \\ 0 & 0 & 1 \end{bmatrix}$$

Therefore, by the definition of T we get $T'(x_*) = 1$. Accordingly, we find $\omega_0(p) = (e-1)p$, $\omega(p) = e^{\frac{1}{e-1}} p$ and $\omega_1(p) = e^{\frac{1}{e-1}}$. Hence, we obtain the radii as $\alpha_1 = 0.382692$ and $\alpha_2 = 0.609316$. Therefore, by definition (13.4), we obtain $\alpha = 0.382692$.

Example 13.4 *Next, we revisit the problem discussed in the introductory section of this paper. $\upsilon = 1$ is a solution. Consequently, we obtain $\omega_0(p) = \omega(p) = 96.8p$ and $\omega_1(p) = 2$. Therefore, the radii are $\alpha_1 = 0.00688705$ and $\alpha_2 = 0.0103757$. Then, by using definition (13.4) we get $\alpha = 0.00688705$.*

13.4 CONCLUSION

The local convergence analysis for the method under consideration is obtained by utilizing generalized Lipschitz condition only on the first derivative. Estimates on computable radii of convergence domains, error bounds and the existence-uniqueness regions for the solution are found. At the end, the suggested theoretical outcomes are verified on application problems. The results obtained are general as it does not depend on the method and can be applied on other methods in a similar manner.

REFERENCES

1. Argyros, C. I., Argyros, I. K., Regmi, S., John, J. A., and Jayaraman, J. (2022). Semi-local convergence of a seventh order method with one parameter for solving non-linear equations. Foundations, 2(4):827–838.
2. Argyros, I. (2007). Computational theory of iterative methods. Elsevier, Oxford.
3. Argyros, I. K. and Hilout, S. (2012). Weaker conditions for the convergence of Newton's method. Journal of Complexity, 28(3):364–387.
4. Argyros, I. K. and Hilout, S. (2013). Computational methods in nonlinear analysis: efficient algorithms, fixed point theory and applications. World Scientific, Singapore.
5. Argyros, I. K. and Kansal, M. (2016). Unified local convergence for a certain family of methods in banach space. SeMA Journal, 73(4):325–334.
6. John, J. A., Jayaraman, J., and Argyros, I. K. (2022). Local convergence of an optimal method of order four for solving non-linear system. International Journal of Applied and Computational Mathematics, 8(4):1–8.
7. Kantorovich, L. V. and Akilov, G. P. (1964). Functional Analysis in Normed Spaces. Pergamon Press, Oxford.
8. Regmi, S., Argyros, I. K., George, S., and Argyros, C. I. (2022). Developments of newton's method under hölder conditions. European Journal of Mathematical Analysis, 2:18–18.
9. Sivakumar, P. and Jayaraman, J. (2019). Some new higher order weighted newton methods for solving nonlinear equation with applications. Mathematical and Computational Applications, 24(2):59.
10. Traub, J. F. (1982). Iterative methods for the solution of equations, volume 312. American Mathematical Soc., Providence.

Index